Maple V

Programming Guide

Springer
New York
Berlin
Heidelberg
Barcelona
Budapest
Hong Kong
London
Milan
Paris
Santa Clara
Singapore
Tokyo

Maple V

Programming Guide

M. B. Monagan   K. O. Geddes   K. M. Heal
G. Labahn   S. M. Vorkoetter

With the Assistance of J. S. Devitt, M. L. Hansen,
D. Redfern, K. M. Rickard

Waterloo Maple Inc.
450 Phillip St.
Waterloo, ON N2L 5J2
Canada

Camera-ready copy prepared using Springer-Verlag's TEX macros.
Printed and bound by Hamilton Printing Company, Rensselaer, NY.
Printed in the United States of America.

9 8 7 6 5 4 3 2 1

ISBN 0-387-98398-8 Springer-Verlag New York Berlin Heidelberg SPIN 10660030
ISBN 0-387-98400-3 Springer-Verlag New York Berlin Heidelberg (Maple V software boxed version) SPIN 10660064

Contents

Introduction

As a Maple user, you may fall into any number of categories. You may only have used Maple interactively. You may already have written many of your own programs. Perhaps, even more fundamentally, you may or may not have programmed in another computer language before attempting your first Maple program. Indeed, you may have used Maple for some time without realizing that the same powerful language you regularly use to enter commands is itself a complete programming language.

Writing a Maple program can be very simple. It may only involve putting a proc() and an end around a sequence of commands that you use every day. On the other hand, the limits for writing Maple procedures with various levels of complexity depend only on you. Over eighty percent of the thousands of commands in the Maple language are themselves Maple programs. You are free to examine these programs and modify them to suit your needs, or extend them so that Maple can tackle new types of problems. You should be able to write useful Maple programs in a few hours, rather than the few days or weeks that it often takes with other languages. This efficiency is partly due to the fact that Maple is *interactive*; this interaction makes it easier to test and correct programs.

Coding in Maple does not require expert programming skills. Unlike traditional programming languages, the Maple language contains many powerful commands which allow you to perform complicated tasks with a single command instead of pages of code. For example, the solve command computes the solution to a system of equations. Maple comes with a large library of prewritten routines, including graphical display primitives, so putting useful programs together from its powerful building blocks is easy.

The aim of this chapter is to provide basic knowledge for proficiently writing Maple code. To learn quickly, read until you encounter some example programs and then write your own variations. This chapter includes many examples along with exercises for you to try. Some of them highlight important differences between Maple and traditional computer languages, which lack symbolic computation capability. Thus, this chapter is also important for those who have written programs in other languages.

This chapter informally presents the most essential elements of the Maple language. You can study the details, exceptions, and options in the other chapters, as the need arises. The examples of basic programming tasks for you to do come with pointers to other chapters and help pages that give further details.

1.1 Getting Started

Maple runs on many different platforms. You can use it through a specialized worksheet interface, or directly through interactive commands typed at a plain terminal. In either case, when you start a Maple session, you will see a Maple prompt character.

>

The prompt character > indicates that Maple is waiting for input.

Throughout this book, the command-line (Maple notation) input format is used. For information on how to toggle between *Maple notation* and *standard math* mode, please refer to the first chapter of the *Learning Guide*.

Your input can be as simple as a single expression. A command is followed immediately by its result.

> 103993/33102;

$$\frac{103993}{33102}$$

Ordinarily, you complete the command with a semicolon and a carriage return. Maple echoes the result—in this case an exact rational number—to the worksheet or the terminal and the particular interface in use, displaying the result as closely to standard mathematical notation as possible.[1]

You may enter commands entirely on one line (as in the previous example) or stretch them across several lines.

---

[1] *Two-Dimensional Expression Output* on page 348 discusses specific commands to control printing.

```
> 103993
> / 33102
> ;
```

$$\frac{103993}{33102}$$

You can even put the terminating semicolon on a separate line. Nothing processes or displays until you complete the command.

Associate names with results by using the assignment statement, :=.

```
> a := 103993/33102;
```

$$a := \frac{103993}{33102}$$

Once assigned a value in this manner, you can use the name a as if it were the value 103993/33102. For example, you can use Maple's evalf command to compute an approximation to 103993/33102 divided by 2.

```
> evalf(a/2);
```

$$1.570796327$$

A Maple *program* is essentially just a prearranged group of commands that Maple always carries out together. The simplest way of creating such a Maple program (or procedure) is to encapsulate the sequence of commands that you would have used to carry out the computation interactively. The following is a program corresponding to the above statement.

```
> half := proc(x)
>      evalf(x/2);
> end;
```

$$half := \mathbf{proc}(x)\ \text{evalf}(1/2 \times x)\ \mathbf{end}$$

The program takes one value, called x within the procedure, and calculates an approximation to that number divided by two. Since this is the last calculation done within the procedure, the half procedure returns this approximation. Give the name half to the procedure using the := notation, just as you would assign a name to any other object. Once you have defined a new procedure, you can use it as a command.

```
> half(2/3);
```

$$.3333333333$$

```
> half(a);
```

$$1.570796327$$

```
> half(1) + half(2);
```

$$1.500000000$$

Merely enclosing the command `evalf(x/2);` between a `proc(...)` and the word end turns it into a procedure.

Create another program corresponding to the following two statements.

```
> a := 103993/33102;

> evalf(a/2);
```

The procedure needs no input.

```
> f := proc()
>     a := 103993/33102;
>     evalf(a/2);
> end;

Warning, 'a' is implicitly declared local
```

$f :=$

proc() **local** a; $a := 103993/33102$; evalf($1/2 \times a$) **end**

Maple's interpretation of this procedure definition appears immediately after the command lines that created it. Examine it carefully and note the following:

- The *name* of this program (procedure) is f.
- The procedure *definition* starts with `proc()`. The empty parentheses indicate that this procedure does not require any *input data*.
- Semicolons separate the individual commands that make up the procedure. Another semicolon after the word end signals the end of the procedure definition.
- You see a display of the procedure definition (just as for any other Maple command) only after you complete it with an end and a semicolon. Even the individual commands that make up the procedure do not display until you complete the entire procedure and enter the last semicolon.
- The *procedure definition* that echoes as the value of the name f is equivalent but not identical to the procedure definition that you entered.
- Maple decided to make your variable a a *local* variable. *Locals and Globals* on page 6 discusses these further. `local` means that the variable a within the procedure is not the same as the variable a outside the

procedure. Thus, it does not matter if you use that name for something else.

Execute the procedure f—that is, cause the statements forming the procedure to execute in sequence—by typing its name followed by parentheses. Enclose any input to the procedure, in this case none, between the parentheses.

```
> f();
```

$$1.570796327$$

You can also refer to the execution of a procedure as an *invocation* or a procedure *call*.

When you invoke a procedure, Maple executes the statements forming *the procedure body* one at a time. The procedure *returns* the result the last statement computed as the *value* of the procedure call.

As with ordinary Maple expressions, you can enter procedure definitions with a large degree of flexibility. Individual statements may appear on different lines, or span several lines. You may also place more than one statement on one line, though that can affect readability of your code. You may even put extra semicolons between statements without causing problems. In some instances, you may omit semicolons.[2]

Sometimes you may not want Maple to display the result of constructing a complicated procedure definition. To suppress the display, use a colon (:) instead of a semicolon (;) at the end of the definition.

```
> g := proc()
>    a := 103993/33102;
>    evalf(a/2);
> end:
```

```
Warning, 'a' is implicitly declared local
```

While a warning about the implicit declaration still occurs, the re-display of the procedure body is suppressed.

Sometimes you may find it necessary to examine the body of a procedure long after constructing it. For ordinary named objects in Maple, such as e, defined below, you can obtain the actual value of the name simply by referring to it by name.

```
> e := 3;
```

$$e := 3$$

---

[2]For example, the semicolon in the definition of a procedure between the last command and the end is optional.

```
> e;
```

$$3$$

If you try this with the procedure g, Maple displays only the name g instead of its true value. Both procedures and tables potentially contain many subobjects. This model of evaluation hides the detail and you can refer to it as *last-name evaluation*. To obtain the true value of the name g, use the eval command, which forces *full evaluation*.

```
> g;
```

$$g$$

```
> eval(g);
```

proc() **local** a; $a := 103993/33102$; evalf$(1/2 \times a)$ **end**

To print the body of a Maple library procedure, set the `interface` variable verboseproc to 2. See `?interface` for details on `interface` variables.

Locals and Globals

Variables that you use at the interactive level in Maple, that is, not within a procedure body, are called *global variables*.

For procedures, you may desire to have variables whose values Maple knows only inside the procedure. These are called *local variables*. While Maple executes a procedure, a global variable by the same name remains unchanged, no matter what value the local variables assume. This allows you to make temporary assignments inside a procedure without affecting anything else in your session.

Scope of variables often refers to the collection of procedures and statements which have access to the value of particular variables. With simple (that is, non-nested) procedures in Maple, only two possibilities exist. Either the value of a name is available everywhere (that is, *global*) or only to the statements that form the particular procedure definition (that is, *local*). The more involved rules that apply for nested procedures are outlined in *Nested Procedures* on page 49.

To demonstrate the distinction between local and global names, first assign values to a global (that is, top level) name b.

```
> b := 2;
```

$$b := 2$$

Next, define two nearly identical procedures: g, explicitly using b as a local variable and h, explicitly using b as a global variable.

```
> g := proc()
>     local b;
>     b := 103993/33102;
>     evalf(b/2);
> end:
```

and

```
> h := proc()
>     global b;
>     b := 103993/33102;
>     evalf(b/2);
> end:
```

Defining the procedures has no effect on the global value of b. In fact, you can even execute the procedure g (which uses local variables) without affecting the value of b.

```
> g();
```

$$1.570796327$$

Therefore, the value of the global variable b is still 2. The procedure g made an assignment to the local variable b which is different from the global variable of the same name.

```
> b;
```

$$2$$

The effect of using the procedure h (which uses *global* variables) is very different.

```
> h();
```

$$1.570796327$$

h changes the global variable b, so it is no longer 2. When you invoke h, the global variable b changes as a *side effect*.

```
> b;
```

$$\frac{103993}{33102}$$

If you do not indicate whether a variable used inside a procedure is local or global, Maple decides on its own and warns you of this. However, you can always use the local or global statements to override Maple's choice. However, it is good programming style to declare all variables either local or global.

Inputs, Parameters, Arguments

An important class of variables that you can use in procedure definitions are neither local nor global. These represent the *inputs* to the procedure. *Parameters* or *arguments* are other names for this class.

Procedure arguments are placeholders for the actual values of data that you supply when you invoke the procedure, which may have more than one argument. The following procedure h accepts *two* quantities, p and q, and constructs the expression p/q.

```
> k := proc(p,q)
>    p/q;
> end:
```

The *arguments* to this procedure are p and q. That is, p and q are place-holders for the actual *inputs* to the procedure.

```
> k(103993,33102);
```

$$\frac{103993}{33102}$$

Maple considers floating-point values to be approximations, rather than exact expressions. If you pass floating-point numbers to a procedure, it returns floating-point numbers.

```
> k( 23, 0.56);
```

$$41.07142857$$

In addition to support for exact and approximate numbers and symbols, Maple provides direct support for *complex* numbers. Maple uses the capital letter I to denote the imaginary unit, $\sqrt{-1}$.

```
> (2 + 3*I)^2;
```

$$-5 + 12\,I$$

```
> k(2 + 3*I, %);
```

$$\frac{2}{13} - \frac{3}{13}\,I$$

```
> k(1.362, 5*I);
```

$$-.2724000000\,I$$

Suppose you want to write a procedure which calculates the norm, $\sqrt{a^2 + b^2}$, of a complex number $z = a + bi$. You can make such a procedure in several ways. The procedure abnorm takes the real and imaginary parts, a and b, as separate input.

```
> abnorm := proc(a,b)
```

```
>     sqrt(a^2+b^2);
> end;
```

$$abnorm := \textbf{proc}(a, b)\, \text{sqrt}(a^2 + b^2)\ \textbf{end}$$

Now abnorm can calculate the norm of $2 + 3i$.

```
> abnorm(2, 3);
```

$$\sqrt{13}$$

You could instead use the Re and Im commands to pick out the *real* and *imaginary* parts, respectively, of a complex number. Hence, you can also calculate the norm of a complex number in the following manner.

```
> znorm := proc(z)
>    sqrt( Re(z)^2 + Im(z)^2 );
> end;
```

$$znorm := \textbf{proc}(z)\, \text{sqrt}(\Re(z)^2 + \Im(z)^2)\ \textbf{end}$$

The norm of $2 + 3i$ is still $\sqrt{13}$.

```
> znorm( 2+3*I );
```

$$\sqrt{13}$$

Finally, you can also compute the norm by re-using the abnorm procedure. The abznorm procedure below uses Re and Im to pass information to abnorm in the form it expects.

```
> abznorm := proc(z)
>    local r, i;
>    r := Re(z);
>    i := Im(z);
>    abnorm(r, i);
> end;
```

$$abznorm :=$$

$$\textbf{proc}(z)\ \textbf{local}\ r,\ i;\ r := \Re(z);\ i := \Im(z);\ abnorm(r, i)\ \textbf{end}$$

Use abznorm to calculate the norm of $2 + 3i$.

```
> abznorm( 2+3*I );
```

$$\sqrt{13}$$

If you do not specify enough information for Maple to calculate the norm, abznorm returns the formula. Here Maple treats x and y as complex numbers. If they were real numbers, then $\Re(x + i\, y)$ would simplify to x.

```
> abznorm( x+y*I );
```

$$\sqrt{\Re(x + I\,y)^2 + \Im(x + I\,y)^2}$$

Many Maple commands return unevaluated in such cases. Thus, you might alter abznorm to return abznorm(x+y*I) in the above example. Later examples in this book show how to give your own procedures this behavior.

1.2   Basic Programming Constructs

This section describes the basic programming constructs you require to get started with real programming tasks. It covers assignment statements, for loops and while loops, conditional statements (if statements), and the use of local and global variables.

The Assignment Statement

Use assignment statements to associate names with computed values. They have the following form.

variable := value ;

This syntax assigns the name on the left-hand side of := to the computed value on the right-hand side. You have seen this statement used in many of the earlier examples.

The use of := here is similar to the assignment statement in programming languages, such as Pascal. Other programming languages, such as C and FORTRAN, use = for assignments. Maple does not use = for assignments, since it is such a natural choice for representing mathematical equations.

Say you want to write a procedure called plotdiff which plots an expression $f(x)$ together with its derivative $f'(x)$ on the interval $[a, b]$. You can accomplish this task by computing the derivative of $f(x)$ with the diff command and then plotting both $f(x)$ and $f'(x)$ on the same interval with the plot command.

```
> y := x^3 - 2*x + 1;
```

$$y := x^3 - 2x + 1$$

Find the derivative of y with respect to x.

```
> yp := diff(y, x);
```

$$yp := 3x^2 - 2$$

Plot *y* and *yp* together.

```
> plot( [y, yp], x=-1..1 );
```

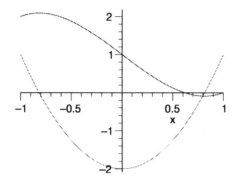

The following procedure combines this sequence of steps.

```
> plotdiff := proc(y,x,a,b)
>       local yp;
>       yp := diff(y,x);
>       plot( [y, yp], x=a..b );
> end;
```

$$plotdiff := \mathbf{proc}(y,\ x,\ a,\ b)$$

$$\mathbf{local}\ yp;$$

$$yp := \mathrm{diff}(y,\ x);\ \mathrm{plot}([y,\ yp],\ x = a..b)$$

$$\mathbf{end}$$

The procedure name is `plotdiff`. It has four parameters: *y*, the expression it will differentiate; *x*, the name of the variable it will use to define the expression; and *a* and *b*, the beginning and the end of the interval over which it will generate the plot. The procedure returns a Maple plot object which you can either display, or use in further plotting routines.

By specifying that *yp* is a local variable, you ensure that its usage in the procedure does not clash with any other usage of the variable that you may have made elsewhere in the current session.

To use the procedure, simply invoke it with appropriate arguments. Plot $\cos(x)$ and its derivative, for *x* running from 0 to 2π.

```
> plotdiff( cos(x), x, 0, 2*Pi );
```

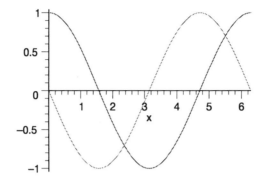

The for Loop

Use looping constructs, such as the for loop, to repeat similar actions a number of times. For example, you can calculate the sum of the first five natural numbers in the following way.

```
> total := 0;
> total := total + 1;
> total := total + 2;
> total := total + 3;
> total := total + 4;
> total := total + 5;
```

You may instead perform the same calculations using a for loop.

```
> total := 0:
> for i from 1 to 5 do
>     total := total + i;
> od;
```

$$total := 1$$
$$total := 3$$
$$total := 6$$
$$total := 10$$
$$total := 15$$

For each cycle through the loop, Maple increments the value of i by one and checks whether i is greater than 5. If it is not, then Maple executes the body of the loop again. When the execution of the loop finishes, the value of `total` is 15.

```
> total;
```

$$15$$

The following procedure uses a `for` loop to calculate the sum of the first n natural numbers.

```
> SUM := proc(n)
>       local i, total;
>       total := 0;
>       for i from 1 to n do
>           total := total+i;
>       od;
>       total;
> end:
```

The purpose of the `total` statement at the end of SUM is to ensure that SUM returns the value `total`. Calculate the sum of the first 100 numbers.

```
> SUM(100);
```

$$5050$$

The `for` statement is an important part of the Maple language, but the language also provides many more succinct and efficient looping constructs.

```
> add(n, n=1..100);
```

$$5050$$

The Conditional Statement

The loop is one of the two most basic constructs in programming. The other basic construct is the `if` or *conditional statement*. It arises in many contexts. For example, you can use the `if` statement to implement an absolute value function.

$$|x| = \begin{cases} x & \text{if } x \geq 0 \\ -x & \text{if } x < 0. \end{cases}$$

Below is a first implementation of ABS. Maple executes the `if` statement as follows: If $x < 0$, then Maple calculates $-x$; otherwise it calculates x. In either case, the absolute value of x is the last result that Maple computes and so is the value that ABS returns.

The closing word `fi` (if in reverse) completes the `if` statement.

```
> ABS := proc(x)
>       if x<0 then
>             -x;
>       else
>              x;
>       fi;
> end;
```

$$ABS := \mathbf{proc}(x) \; \mathbf{if} \; x < 0 \; \mathbf{then} \; -x \; \mathbf{else} \; x \; \mathbf{fi} \; \mathbf{end}$$

```
> ABS(3); ABS(-2.3);
```

$$3$$

$$2.3$$

Returning Unevaluated   The ABS procedure above cannot handle non-numeric input.

```
> ABS( a );
```

```
Error, (in ABS) cannot evaluate boolean
```

The problem is that since Maple knows nothing about a, it cannot determine whether a is less than zero. In such cases, your procedure should *return unevaluated*; that is, ABS should return ABS(a).

```
> 'ABS'(a);
```

$$ABS(a)$$

The single quotes tell Maple not to evaluate ABS. You can modify the ABS procedure using the `type(..., numeric)` command to test whether x is a number.

```
> ABS := proc(x)
>    if type(x,numeric) then
>         if x<0 then -x else x fi;
>    else
>          'ABS'(x);
>    fi;
> end:
```

The above ABS procedure contains an example of a *nested* `if` statement, that is, one `if` statement appearing within another. You need an even more complicated nested `if` statement to implement the function

$$\text{hat}(x) = \begin{cases} 0 & \text{if } x \leq 0 \\ x & \text{if } 0 < x \leq 1 \\ 2 - x & \text{if } 1 < x \leq 2 \\ 0 & \text{if } x > 2. \end{cases}$$

Here is a first version of HAT.

```
> HAT := proc(x)
>    if type(x, numeric) then
>       if x<=0 then
>          0;
>       else
>          if x<=1 then
>             x;
>          else
>             if x<=2 then
>                2-x;
>             else
>                0;
>             fi;
>          fi;
>       fi;
>    else
>       'HAT'(x);
>    fi;
> end:
```

The indentations make it easier to identify which statements belong to which `if` conditions. Even so, a better approach is to use the optional `elif` clause (else if) in the second-level `if` statement.

```
> HAT := proc(x)
>    if type(x, numeric) then
>       if x<=0 then 0;
>       elif x<=1 then x;
>       elif x<=2 then 2-x;
>       else 0;
>       fi;
>    else
>       'HAT'(x);
>    fi;
> end:
```

You may use as many `elif` branches as you need.

Symbolic Transformations You can improve the ABS procedure from the last section even further. Consider the product *ab*. Since *ab* is an unknown, ABS returns unevaluated.

```
> ABS( a*b );
```

$$\text{ABS}(a\,b)$$

However, the absolute value of a product is the product of the absolute values.

$$|ab| \rightarrow |a||b|.$$

That is, ABS should map over products.

```
> map( ABS, a*b );
```

$$\text{ABS}(a)\,\text{ABS}(b)$$

You can use the type(..., '*') command to test whether an expression is a product and use the map command to apply ABS to each operand of the product.

```
> ABS := proc(x)
>     if type(x, numeric) then
>         if x<0 then -x else x fi;
>     elif type(x, '*') then
>         map(ABS, x);
>     else
>         'ABS'(x);
>     fi;
> end:
> ABS( a*b );
```

$$\text{ABS}(a)\,\text{ABS}(b)$$

This feature is especially useful if some of the factors are numbers.

```
> ABS( -2*a );
```

$$2\,\text{ABS}(a)$$

You may want to improve ABS further so that it can calculate the absolute value of a complex number.

Parameter Type Checking Sometimes when you write a procedure, you intend it to handle only a certain type of input. Calling the procedure with a different type of input may not make any sense. You can use type checking to verify that the inputs to your procedure are of the correct type. Type checking is especially important for complicated procedures as it helps you to identify mistakes.

Consider the original implementation of SUM.

```
> SUM := proc(n)
>     local i, total;
>     total := 0;
>     for i from 1 to n do
>         total := total+i;
>     od;
>     total;
> end:
```

Clearly, *n* should be an integer. If you try to use the procedure on symbolic data, it breaks.

```
> SUM("hello world");

Error, (in SUM)
final value in for loop must be numeric or character
```

The error message indicates what went wrong inside the `for` statement while trying to execute the procedure. The test in the `for` loop failed because "hello world" is a string, not a number, and Maple could not determine whether to execute the loop. The following implementation of SUM provides a much more informative error message. The `type(...,integer)` command determines whether *n* is an integer.

```
> SUM := proc(n)
>     local i,total;
>     if not type(n, integer) then
>         ERROR("input must be an integer");
>     fi;
>     total := 0;
>     for i from 1 to n do  total := total+i  od;
>     total;
> end:
```

Now the error message is more helpful.

```
> SUM("hello world");

Error, (in SUM) input must be an integer
```

Using type to check inputs is such a common task that Maple provides a simple means of declaring the type of an argument to a procedure. For example, you can rewrite the SUM procedure in the following manner. An informative error message helps you to find and correct a mistake quickly.

```
> SUM := proc(n::integer)
>    local i, total;
>    total := 0;
>    for i from 1 to n do  total := total+i  od;
>    total;
> end:

> SUM("hello world");

Error, SUM expects its 1st argument, n, to be of type
integer, but received "hello world"
```

Maple understands a large number of types. In addition, you can combine existing types algebraically to form new types, or you can define entirely new types. See ?type.

The while Loop

The while loop is an important type of structure. It has the following structure.

> while *condition* do *commands* od;

Maple tests the *condition* and executes the *commands* inside the loop over and over again until the *condition* fails.

You can use the while loop to write a procedure that divides an integer n by two as many times as is possible. The iquo and irem commands calculate the quotient and remainder, respectively, using integer division.

```
> iquo( 8, 3 );
```

$$2$$

```
> irem( 8, 3 );
```

$$2$$

Thus, you can write a divideby2 procedure in the following manner.

```
> divideby2 := proc(n::posint)
>    local q;
>    q := n;
>    while irem(q, 2) = 0 do
>        q := iquo(q, 2);
>    od;
>    q;
> end:
```

Apply divideby2 to 32 and 48.

```
> divideby2(32);
```

$$1$$

```
> divideby2(48);
```

$$3$$

The while and for loops are both special cases of a more general repetition statement; see *The Repetition Statement* on page 125.

Modularization

When you write procedures, identifying subtasks and writing these as separate procedures is a good idea. Doing so makes your procedures easier to read, and you may be able to reuse some of the subtask procedures in another application.

Consider the following mathematical problem: Suppose you have a positive integer, say forty.

```
> 40;
```

$$40$$

Divide the integer by two, as many times as possible; the divideby2 procedure above does just that for you.

```
> divideby2( % );
```

$$5$$

Multiply the result by three and add one.

```
> 3*% + 1;
```

$$16$$

Again, divide by two.

```
> divideby2( % );
```

$$1$$

Multiply by three and add one.

```
> 3*% + 1;
```

$$4$$

Divide.

```
> divideby2( % );
```

$$1$$

The result is 1 again, so from now on you will get 4, 1, 4, 1, ... Mathematicians have conjectured that you always reach the number 1 in this way, no matter which positive integer you begin with. You can study this conjecture, known as *the 3n + 1 conjecture*, by writing a procedure which calculates how many iterations you need to get to the number 1. The following procedure makes a single iteration.

```
> iteration := proc(n::posint)
>     local a;
>     a := 3*n + 1;
>     divideby2( a );
> end:
```

The checkconjecture procedure counts the number of iterations.

```
> checkconjecture := proc(x::posint)
>     local count, n;
>     count := 0;
>     n := divideby2(x);
>     while n>1 do
>         n := iteration(n);
>         count := count + 1;
>     od;
>     count;
> end:
```

You can now check the conjecture for different values of *x*.

```
> checkconjecture( 40 );
```

$$1$$

```
> checkconjecture( 4387 );
```

$$49$$

You could write checkconjecture as one self-contained procedure without references to iteration or divideby2. But then, you would have to use nested while statements, thus making the procedure much harder to read.

Recursive Procedures

Just as you can write procedures that call other procedures, you can also write a procedure that calls itself. This is called *recursive programming*. As an example, consider the Fibonacci numbers which the following defines.

$$f_n = f_{n-1} + f_{n-2} \qquad \text{for } n \geq 2,$$

where $f_0 = 0$, and $f_1 = 1$. The following procedure calculates f_n for any n.

```
> Fibonacci := proc(n::nonnegint)
>    if n<2 then
>       n;
>    else
>       Fibonacci(n-1)+Fibonacci(n-2);
>    fi;
> end:
```

Here is a sequence of the first sixteen Fibonacci numbers.

```
> seq( Fibonacci(i), i=0..15 );
```

$$0, 1, 1, 2, 3, 5, 8, 13, 21, 34, 55, 89, 144, 233, 377, 610$$

The `time` command tells you the number of seconds a procedure takes to execute. Fibonacci is not very efficient.

```
> time( Fibonacci(20) );
```

$$5.794$$

The reason is that `Fibonacci` recalculates the same results over and over again. To find f_{20}, it must find f_{19} and f_{18}; to find f_{19}, it must find f_{18} again and f_{17}; and so on. One solution to this efficiency problem is to tell `Fibonacci` to remember its results. That way, `Fibonacci` only has to calculate f_{18} once. The `remember` option makes a procedure store its results in a *remember table*. *Remember Tables* on page 68 further discusses remember tables.

```
> Fibonacci := proc(n::nonnegint)
>    option remember;
>    if n<2 then
>       n;
>    else
>       Fibonacci(n-1)+Fibonacci(n-2);
>    fi;
> end:
```

This version of `Fibonacci` is much faster.

```
> time( Fibonacci(20) );
```

$$.006$$

```
> time( Fibonacci(2000) );
```

$$.133$$

If you use remember tables indiscriminately, Maple may run out of memory. You can often rewrite recursive procedures using a loop, but recursive procedures are usually easier to read. The procedure below is a loop version of Fibonacci.

```
> Fibonacci := proc(n::nonnegint)
>     local temp, fnew, fold, i;
>     if n<2 then
>         n;
>     else
>         fold := 0;
>         fnew := 1;
>         for i from 2 to n do
>             temp := fnew + fold;
>             fold := fnew;
>             fnew := temp;
>         od;
>         fnew;
>     fi;
> end:

> time( Fibonacci(2000) );
```

$$.133$$

When you write recursive procedures, you must weigh the benefits of remember tables against their use of memory. Also, you must make sure that your recursion stops.

The RETURN Command   A Maple procedure by default returns the result of the last computation within the procedure. You can use the RETURN command to override this behavior. In the version of Fibonacci below, if $n < 2$ then the procedure returns n and Maple does not execute the rest of the procedure.

```
> Fibonacci := proc(n::nonnegint)
>     option remember;
>     if n<2 then
>         RETURN(n);
>     fi;
>     Fibonacci(n-1)+Fibonacci(n-2);
> end:
```

Using the RETURN command can make your recursive procedures easier to read; the usually complicated code that handles the general step of the recursion does not end up inside a nested if statement.

Exercise

1. The Fibonacci numbers satisfy the following recurrence.

$$F(2n) = 2F(n-1)F(n) + F(n)^2 \quad \text{where } n > 1$$

and

$$F(2n+1) = F(n+1)^2 + F(n)^2 \quad \text{where } n > 1$$

Use these new relations to write a recursive Maple procedure which computes the Fibonacci numbers. How much recomputation does this procedure do?

1.3 Basic Data Structures

The programs developed so far in this chapter have operated primarily on a single number or a single formula. More advanced programs often manipulate more complicated collections of data. A *data structure* is a systematic way of organizing data. The organization you choose for your data can directly affect the style of your programs and how fast they execute.

Maple has a rich set of built-in data structures. This section will address *sequences*, *lists*, and *sets*.

Many Maple commands take sequences, lists, and sets as inputs, and produce sequences, lists, and sets as outputs. Here are some examples of how such data structures are useful in solving problems.

PROBLEM: Write a Maple procedure which given $n > 0$ data values $x_1, x_2, \ldots, x_n$ computes their average, where the following equation gives the average of n numbers.

$$\mu = \frac{1}{n} \sum_{i=1}^{n} x_i.$$

You can easily represent the data for this problem as a list. nops gives the total number of entries in a list X, while the ith entry of the list is X[i].

```
> X := [1.3, 5.3, 11.2, 2.1, 2.1];
```
$$X := [1.3, 5.3, 11.2, 2.1, 2.1]$$

```
> nops(X);
```

$$5$$

```
> X[2];
```

$$5.3$$

The most efficient way to add the numbers in a list is to use the add command.

```
> add( i, i=X );
```

$$22.0$$

The procedure average below computes the average of the entries in a list. It handles empty lists as a special case.

```
> average := proc(X::list)
>    local n, i, total;
>    n := nops(X);
>    if n=0 then ERROR("empty list") fi;
>    total := add(i, i=X);
>    total / n;
> end:
```

Using this procedure you can find the average of the list X.

```
> average(X);
```

$$4.400000000$$

The procedure still works if the list has symbolic entries.

```
> average( [ a , b , c ] );
```

$$\frac{1}{3}a + \frac{1}{3}b + \frac{1}{3}c$$

Exercise

1. Write a Maple procedure called sigma which, given $n > 1$ data values, $x_1, x_2, \ldots, x_n$, computes their standard deviation. The following equation gives the standard deviation of $n > 1$ numbers,

$$\sigma = \sqrt{\frac{1}{n}\sum_{i=1}^{n}(x_i - \mu)^2}$$

where μ is the average of the data values.

You create lists and many other objects in Maple out of more primitive data structures called *sequences*. The list X above contains the following sequence.

```
> Y := X[];
```

$$Y := 1.3,\ 5.3,\ 11.2,\ 2.1,\ 2.1$$

You can select elements from a sequence in the same way you select elements from a list.

```
> Y[3];
```

$$11.2$$

```
> Y[2..4];
```

$$5.3,\ 11.2,\ 2.1$$

The important difference between sequences and lists is that Maple flattens a sequence of sequences into a single sequence.

```
> W := a,b,c;
```

$$W := a,\ b,\ c$$

```
> Y, W, Y;
```

$$1.3,\ 5.3,\ 11.2,\ 2.1,\ 2.1,\ a,\ b,\ c,\ 1.3,\ 5.3,\ 11.2,\ 2.1,\ 2.1$$

In contrast, a list of lists remains just that, a list of lists.

```
> [ X, [a,b,c], X ];
```

$$[[1.3,\ 5.3,\ 11.2,\ 2.1,\ 2.1],\ [a,\ b,\ c],$$
$$[1.3,\ 5.3,\ 11.2,\ 2.1,\ 2.1]]$$

If you enclose a sequence in a pair of braces, you get a *set*.

```
> Z := { Y };
```

$$Z := \{1.3,\ 5.3,\ 2.1,\ 11.2\}$$

As in mathematics, a set is an *unordered* collection of distinct objects. Hence, Z has only four elements as the nops command demonstrates.

```
> nops(Z);
```

$$4$$

You can select elements from a set in the same way you select elements from a list or a sequence, but the order of the elements in a set is session dependent. Do not make any assumptions about this order.

You may also use the `seq` command to build sequences.

```
> seq( i^2, i=1..5 );
```

$$1, 4, 9, 16, 25$$

```
> seq( f(i), i=X );
```

$$f(1.3), f(5.3), f(11.2), f(2.1), f(2.1)$$

You can create lists or sets by enclosing a sequence in square brackets or braces, respectively. The following command creates a list of sets.

```
> [ seq( { seq( i^j, j=1..3) }, i=-2..2 ) ];
```

$$[\{-2, 4, -8\}, \{-1, 1\}, \{0\}, \{1\}, \{2, 4, 8\}]$$

You can also create a sequence using a loop. NULL is the empty sequence.

```
> s := NULL;
```

$$s :=$$

```
> for i from 1 to 5 do
>     s := s, i^2;
> od;
```

$$s := 1$$
$$s := 1, 4$$
$$s := 1, 4, 9$$
$$s := 1, 4, 9, 16$$
$$s := 1, 4, 9, 16, 25$$

However, the `for` loop is much less efficient as it must execute many statements and create all the intermediate sequences. The seq command creates the sequence in one step.

Exercise

1. Write a Maple procedure which, given a list of lists of numerical data, computes the means of each column of the data.

A MEMBER Procedure

You may want to write a procedure that determines whether a certain object is an element of a list or a set. The procedure below uses the RETURN command discussed in *The RETURN Command* on page 22.

```
> MEMBER := proc( a::anything, L::{list, set} )
>     local i;
>     for i from 1 to nops(L) do
>         if a=L[i] then RETURN(true) fi;
>     od;
>     false;
> end:
```

Here 3 is a member of the list.

```
> MEMBER( 3, [1,2,3,4,5,6] );
```

true

The type of loop that MEMBER uses occurs so frequently that Maple has a special version of the for loop for it.

```
> MEMBER := proc( a::anything, L::{list, set} )
>     local i;
>     for i in L do
>         if a=i then RETURN(true) fi;
>     od;
>     false;
> end:
```

The symbol x is not a member of this set.

```
> MEMBER( x, {1,2,3,4} );
```

false

Instead of using your own MEMBER procedure, you can use the built-in member command.

Exercise

1. Write a Maple procedure called POSITION which returns the position i of an element x in a list L. That is, POSITION(x,L) should return an integer $i > 0$ such that L[i]=x. Return 0 if x is not in the list L.

Binary Search

One of the most basic and well studied computing problems is that of searching. A typical problem involves searching a list of words (a dictionary, for example) for a specific word w.

Many possible solutions are available. One approach is to search the list by comparing each word in turn with w until Maple either finds w or it reaches the end of the list.

```
> Search := proc(Dictionary::list, w::anything)
>     local x;
>     for x in Dictionary do
>         if x=w then RETURN(true) fi;
>     od;
>     false
> end:
```

However, if the Dictionary is large, say 50 000 entries, this approach can take a long time.

You can reduce the execution time required by sorting the Dictionary before you search it. If you sort the dictionary into ascending order then you can stop searching as soon as you encounter a word *greater than w*. On average, you only have to look halfway through the dictionary.

Binary searching provides an even better approach. Check the word in the middle of the dictionary. Since you already sorted the dictionary you can tell whether w is in the first or the second half. Repeat the process with the appropriate half of the dictionary. The procedure below searches the dictionary, D, for the word, w, from position, s, to position, f, in D. The lexorder command determines the lexicographical ordering of two strings.

```
> BinarySearch :=
> proc(D::list(string), w::string, s::integer, f::integer)
>     local m;
>     if s>f then RETURN(false) fi; # entry was not found.
>     m := iquo(s+f+1, 2);  # midpoint of D.
>     if w=D[m] then
>         true;
>     elif lexorder(w, D[m]) then
>         BinarySearch(D, w, s, m-1);
>     else
>         BinarySearch(D, w, m+1, f);
>     fi;
> end:
```

Here is a short dictionary.

```
> Dictionary := [ "induna", "ion", "logarithm", "meld" ];
```

$$Dictionary := [\text{“induna”}, \text{“ion”}, \text{“logarithm”}, \text{“meld”}]$$

Now search the dictionary for a few words.

```
> BinarySearch( Dictionary, "hedgehogs", 1, nops(Dictionary) );
```

false

```
> BinarySearch( Dictionary, "logarithm", 1, nops(Dictionary) );
```

$$true$$

```
> BinarySearch( Dictionary, "melodious", 1, nops(Dictionary) );
```

$$false$$

Exercises

1. Can you demonstrate that the BinarySearch procedure always terminates? Suppose the dictionary has n entries. How many words in the dictionary D does BinarySearch look at in the worst case?
2. Recode BinarySearch to use a while loop instead of calling itself recursively.

Plotting the Roots of a Polynomial

You can construct lists of any type of object, even lists. A list of two numbers often represents a point in the plane. The plot command uses this structure to generate plots of points and lines.

```
> plot( [ [ 0, 0], [ 1, 2], [-1, 2] ],
>     style=point, color=black );
```

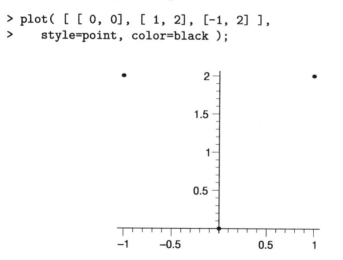

You can use this approach to write a procedure which plots the complex roots of a polynomial. Consider the polynomial $x^3 - 1$.

```
> y := x^3-1;
```

$$y := x^3 - 1$$

Numeric solutions are sufficient for plotting.

```
> R := [ fsolve(y=0, x, complex) ];
```

$$R := [-.5000000000 - .8660254038\,I,$$

$$-.5000000000 + .8660254038\,I,\ 1.]$$

You need to turn this list of complex numbers into a list of points in the plane. The Re and Im commands pick out the real and imaginary parts, respectively.

```
> points := map( z -> [Re(z), Im(z)], R );
```

$$points := [[-.5000000000, -.8660254038],$$

$$[-.5000000000, .8660254038],\ [1., 0]]$$

You can now plot the points.

```
> plot( points, style=point);
```

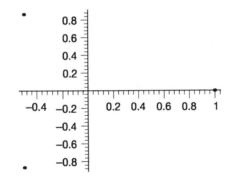

You can automate this technique. The input should be a polynomial in x with constant coefficients.

```
> rootplot := proc( p::polynom(constant, x) )
>    local R, points;
>    R := [ fsolve(p, x, complex) ];
>    points := map( z -> [Re(z), Im(z)], R );
>    plot( points, style=point );
> end:
```

Here is a plot of the roots of the polynomial $x^6 + 3x^5 + 5x + 10$.

```
> rootplot( x^6+3*x^5+5*x+10 );
```

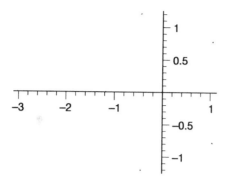

The `randpoly` command generates a random polynomial.

```
> y := randpoly(x, degree=100);
```

$$y := 79\,x^{44} + 56\,x^{30} + 49\,x^{24} + 63\,x^{18} + 57\,x^{71} - 59\,x^{63}$$

```
> rootplot( y );
```

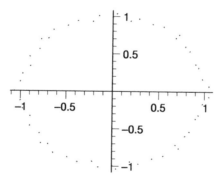

When you write procedures, you often have several choices of how to represent the data that your procedures work with. The choice of data structure can have great impact on how easy it is to write your procedure and how efficient it is. *Choosing a Data Structure: Connected Graphs* on page 63 describes an example of choosing a data structure.

1.4 Computing with Formulae

Maple's real strength stems from its ability to perform symbolic manipulations. This section demonstrates some of these capabilities through sample programs for computation with polynomials. While the examples are specific to polynomials, the techniques and methods apply to more general formulae.

In mathematics, a *polynomial* in the single variable, x, is most easily recognizable in the expanded form,

$$\sum_{i=0}^{n} a_i x^i, \qquad \text{where if } n > 0, \text{ then } a_n \neq 0.$$

The a_is are the *coefficients*. They can be numbers or even expressions involving variables. The crucial point is that each coefficient is independent of (does not contain) x.

The Height of a Polynomial

The *height* of a polynomial is the magnitude (absolute value) of the largest coefficient. The procedure below finds the height of a polynomial, p, in the variable x. The degree command finds the degree of a polynomial and the coeff command extracts specific coefficients from a polynomial.

```
> HGHT := proc(p::polynom, x::name)
>    local i, c, height;
>    height := 0;
>    for i from 0 to degree(p, x) do
>       c := coeff(p, x, i);
>       height := max(height, abs(c));
>    od;
>    height;
> end:
```

The height of $32x^6 - 48x^4 + 18x^2 - 1$ is 48.

```
> p := 32*x^6-48*x^4+18*x^2-1;
```

$$p := 32\,x^6 - 48\,x^4 + 18\,x^2 - 1$$

```
> HGHT(p,x);
```

$$48$$

A significant weakness of the HGHT procedure is its inefficiency with sparse polynomials; that is, polynomials with few terms relative to their degree. For example, to find the height of $x^{4321} - 1$ the HGHT procedure has to examine 4322 coefficients.

The coeffs command returns the sequence of coefficients of a polynomial.

```
> coeffs( p, x );
```

$$-1,\ 32,\ -48,\ 18$$

You cannot map the abs command, or any other command, onto a sequence. One solution is to turn the sequence into a list or a set.

```
> S := map( abs, {%} );
```

$$S := \{1,\ 18,\ 32,\ 48\}$$

The max command, however, works on sequences, so now you must turn the set into a sequence again.

```
> max( S[] );
```

$$48$$

The following version of HGHT uses this technique.

```
> HGHT := proc(p::polynom, x::name)
>     local S;
>     S := { coeffs(p, x) };
>     S := map( abs, S );
>     max( S[] );
> end:
```

Try the procedure out on a random polynomial.

```
> p := randpoly(x, degree=100 );
```

$$p := 79\,x^{71} + 56\,x^{63} + 49\,x^{44} + 63\,x^{30} + 57\,x^{24} - 59\,x^{18}$$

```
> HGHT(p, x);
```

$$79$$

If the polynomial is in expanded form, you can also find its height in the following manner. You can map a command directly onto a polynomial. The map command applies the command to each term in the polynomial.

```
> map( f, p );
```

$$f(79\,x^{71}) + f(56\,x^{63}) + f(49\,x^{44}) + f(63\,x^{30}) + f(57\,x^{24})$$
$$+ f(-59\,x^{18})$$

Thus, you can map abs directly onto the polynomial.

```
> map( abs, p );
```

$$79\,|x|^{71} + 56\,|x|^{63} + 49\,|x|^{44} + 63\,|x|^{30} + 57\,|x|^{24}$$
$$+ 59\,|x|^{18}$$

Then use `coeffs` to find the sequence of coefficients of that polynomial.

```
> coeffs( % );
```

$$79,\ 56,\ 49,\ 63,\ 57,\ 59$$

Finally, find the maximum.

```
> max( % );
```

$$79$$

Hence, you can calculate the height of a polynomial with this one-liner.

```
> p := randpoly(x, degree=50) * randpoly(x, degree=99);
```

$$p := (77\,x^{48} + 66\,x^{44} + 54\,x^{37} - 5\,x^{20} + 99\,x^5 - 61\,x^3)$$
$$(-47\,x^{57} - 91\,x^{33} - 47\,x^{26} - 61\,x^{25} + 41\,x^{18} - 58\,x^8)$$

```
> max( coeffs( map(abs, expand(p)) ) );
```

$$9214$$

Exercise

1. Write a procedure which computes the Euclidean norm of a polynomial; that is, $\sqrt{\sum_{i=0}^{n} |a_i|^2}$.

The Chebyshev Polynomials, $T_n(x)$

The Chebyshev polynomials, $T_n(x)$, satisfy the following linear recurrence.

$$T_n(x) = 2x\,T_{n-1}(x) + T_{n-2}(x), \qquad \text{for } n \geq 2.$$

The first two Chebyshev polynomials are $T_0(x) = 1$ and $T_1(x) = x$. This example is similar to the Fibonacci example in *Recursive Procedures* on page 20. Here is a simple procedure, T, that computes $T_n(x)$.

```
> T := proc(n::nonnegint, x::name)
>    option remember;
>    if n=0 then
>        RETURN(1);
>    elif n=1 then
>        RETURN(x);
>    fi;
>    2*x*T(n-1,x) - T(n-2,x);
```

```
> end:
```

Maple does not automatically expand the polynomial.

```
> T(4,x);
```

$$2x(2x(2x^2 - 1) - x) - 2x^2 + 1$$

You can expand the polynomial yourself.

```
> expand(%);
```

$$8x^4 - 8x^2 + 1$$

You may be tempted to rewrite the procedure so that it expands the result before returning it. However, this may be a waste of effort since you do not know whether or not the user of your procedure wants the Chebyshev polynomial in expanded form. Moreover, since the T procedure is recursive, it would expand all the intermediate results as well.

Exercise

1. The Fibonacci polynomials, $F_n(x)$, satisfy the linear recurrence

$$F_n(x) = x F_{n-1}(x) + F_{n-2}(x),$$

where $F_0(x) = 0$ and $F_1(x) = 1$. Write a Maple procedure to compute and factor $F_n(x)$. Can you see any pattern?

Integration by Parts

Maple's indefinite integral evaluator is very powerful. This section describes how you could write your own procedure for integrating formulae of the form

$$p(x)f(x),$$

where $p(x)$ is a polynomial in x and $f(x)$ is a special function. Here $p(x) = x^2$ and $f(x) = e^x$.

```
> int( x^2*exp(x), x );
```

$$x^2 e^x - 2x e^x + 2e^x$$

As another example, here $p(x) = x^3$ and $f(x) = \sin^{-1}(x)$.

```
> int( x^3*arcsin(x), x );
```

$$\frac{1}{4} x^4 \arcsin(x) + \frac{1}{16} x^3 \sqrt{1 - x^2} + \frac{3}{32} x \sqrt{1 - x^2}$$
$$- \frac{3}{32} \arcsin(x)$$

Usually you would use *integration by parts* to compute integrals of this form.

```
> int( u(x)*v(x), x ) = u(x)*int(v(x),x) -
> int( diff(u(x),x) * int(v(x),x), x );
```

$$\int u(x)\,v(x)\,dx = u(x) \int v(x)\,dx - \int \left(\frac{\partial}{\partial x} u(x) \right) \int v(x)\,dx\,dx$$

You can verify this formula by differentiating both sides of the equation.

```
> diff(%,x);
```

$$u(x)\,v(x) = u(x)\,v(x)$$

```
> evalb(%);
```

$$true$$

Applying integration by parts to the first example yields

$$\int x^n e^x\,dx = x^n \int e^x\,dx - \int (nx^{n-1} \int e^x\,dx)\,dx = x^n e^x - n \int x^{n-1} e^x\,dx.$$

It introduces a new integral, but the degree of x in that new integral is one smaller than in the old integral. By applying the formula repeatedly, the problem eventually reduces to evaluating $\int e^x$, which is simply e^x.

The following procedure uses integration by parts to calculate the integral

$$\int x^n e^x\,dx,$$

by calling itself recursively until $n = 0$.

```
> IntExpMonomial := proc(n::nonnegint, x::name)
>     if n=0 then RETURN( exp(x) ) fi;
>     x^n*exp(x) - n*IntExpMonomial(n-1, x);
> end:
```

IntExpMonomial can calculate $\int x^5 e^x\,dx$.

```
> IntExpMonomial(5, x);
```

$$x^5 e^x - 5 x^4 e^x + 20 x^3 e^x - 60 x^2 e^x + 120 x e^x - 120 e^x$$

You can simplify this answer using the `collect` command to group the terms involving $\exp(x)$ together.

```
> collect(%, exp(x));
```

$$(x^5 - 5 x^4 + 20 x^3 - 60 x^2 + 120 x - 120) e^x$$

You can now write a procedure which calculates $\int p(x)e^x \, dx$ for any polynomial p. The idea is that integration is linear:

$$\int af(x) + g(x) \, dx = a \int f(x) \, dx + \int g(x) \, dx.$$

The `IntExpPolynomial` procedure below uses `coeff` to extract the coefficients of p one at a time.

```
> IntExpPolynomial := proc(p::polynom, x::name)
>     local i, result;
>     result := add( coeff(p, x, i)*IntExpMonomial(i, x),
>                     i=0..degree(p, x) );
>     collect(result, exp(x));
> end:
```

Here `IntExpPolynomial` calculates $\int (x^2 + 1)(1 - 3x)e^x \, dx$.

```
> IntExpPolynomial( (x^2+1)*(1-3*x), x );
```

$$(24 - 23 x + 10 x^2 - 3 x^3) e^x$$

Exercise

1. Modify the procedure `IntExpPolynomial` to be more efficient by processing only the non-zero coefficients of $p(x)$.

Computing with Symbolic Parameters

The polynomial $2x^5 + 1$ is an example of an *explicit polynomial* in x. All the elements of the polynomial, except x, are explicit numbers. On the other hand, polynomials like $3x^n + 2$, where n is an unspecified positive integer, or $a + x^5$, where a is an unknown which is independent of x, are examples of *symbolic polynomials*; they contain additional unspecified symbolic parameters.

The procedure `IntExpPolynomial` in *Integration by Parts* on page 35 calculates the integral $\int p(x)e^x \, dx$ where p is an explicit polynomial. In its present version `IntExpPolynomial` cannot handle symbolic polynomials.

```
> IntExpPolynomial( a*x^n, x );
```

```
Error, IntExpPolynomial expects its 1st argument, p,
to be of type polynom, but received a*x^n
```

You may want to extend `IntExpPolynomial` so that it can integrate $p(x)e^x$ for symbolic polynomials p as well. The first problem is that of finding a formula for $\int x^n e^x \, dx$ for any natural number n. Often you can find such a formula by carefully examining the pattern for specific results. Here are the first few results for explicit values of n.

```
> IntExpPolynomial(x, x);
```

$$(x - 1)\, e^x$$

```
> IntExpPolynomial(x^2, x);
```

$$(x^2 - 2\,x + 2)\, e^x$$

```
> IntExpPolynomial(x^3, x);
```

$$(x^3 - 3\,x^2 + 6\,x - 6)\, e^x$$

With sufficient time and ingenuity you would find the formula

$$\int x^n e^x \, dx = n!\, e^x \sum_{i=0}^{n} \frac{(-1)^{n-i} x^i}{i!}.$$

This formula holds only for non-negative integers n. Use the *assume* facility to tell Maple that the unknown n has certain properties.

```
> assume(n, integer);
> additionally(n >= 0);
```

Note that a simple type check is not sufficient to determine that n is an integer.

```
> type(n, integer);
```

false

You need to use the `is` command, which is part of the assume facility.

```
> is(n, integer), is(n >= 0);
```

true, true

Thus, you can rewrite the `IntExpMonomial` procedure from *Integration by Parts* on page 35 in the following manner.

```
> IntExpMonomial := proc(n::anything, x::name)
>    local i;
>    if is(n, integer) and is(n >= 0) then
>        n! * exp(x) * sum( ( (-1)^(n-i)*x^i )/i!, i=0..n );
>    else
>        ERROR("Expected a non-negative integer but recieved", n);
>    fi;
> end:
```

This version of IntExpMonomial accepts both explicit and symbolic input.

```
> IntExpMonomial(4, x);
```

$$24\, e^x \left(1 - x + \frac{1}{2}x^2 - \frac{1}{6}x^3 + \frac{1}{24}x^4\right)$$

In the next example, Maple evaluates the sum in terms of the gamma function. The tilde (~) on n indicates that n carries an assumption.

```
> IntExpMonomial(n, x);
```

$$n!\, e^x ((-1)^{n\tilde{}}\, e^{(-x)} + x^{(n\tilde{}+1)}(
$$
$$-(n\tilde{}+1)\,(-x)^{(-1-n\tilde{})}\, e^{(-x)}\, \Gamma(n\tilde{}+1,\, -x)
$$
$$+(-x)^{(-1-n\tilde{})}\, e^{(-x)}\, \Gamma(2+n\tilde{}))/(n\tilde{}+1)!)$$

You can check the answer by differentiating it with respect to x. The simplify command reveals $x^n e^x$ as expected.

```
> diff(%, x);
```

$$n!\, e^x ((-1)^{n\tilde{}}\, e^{(-x)}
$$
$$+\frac{x^{(n\tilde{}+1)}\,(-\%1 + (-x)^{(-1-n\tilde{})}\, e^{(-x)}\, \Gamma(2+n\tilde{}))}{(n\tilde{}+1)!}) + n!
$$
$$e^x(-(-1)^{n\tilde{}}\, e^{(-x)} + x^{(n\tilde{}+1)}\,(n\tilde{}+1)
$$
$$(-\%1 + (-x)^{(-1-n\tilde{})}\, e^{(-x)}\, \Gamma(2+n\tilde{}))/(x\,(n\tilde{}+1)!)+
$$
$$x^{(n\tilde{}+1)}(
$$
$$-(n\tilde{}+1)\,(-x)^{(-1-n\tilde{})}\,(-1-n\tilde{})\, e^{(-x)}\, \Gamma(n\tilde{}+1,\, -x)/x
$$
$$+\%1 - (n\tilde{}+1)\,(-x)^{(-1-n\tilde{})}\, e^{(-x)}\,(-x)^{n\tilde{}}\, e^x
$$
$$+\frac{(-x)^{(-1-n\tilde{})}\,(-1-n\tilde{})\, e^{(-x)}\, \Gamma(2+n\tilde{})}{x}$$

$$-(-x)^{(-1-n\tilde{})}e^{(-x)}\,\Gamma(2+n\tilde{}))/(n\tilde{}+1)!)$$

$$\%1 := (n\tilde{}+1)\,(-x)^{(-1-n\tilde{})}e^{(-x)}\,\Gamma(n\tilde{}+1,\,-x)$$

> simplify(%);

$$x^{n\tilde{}}\,e^{x}$$

Clearly, the use of symbolic constants in this way greatly extends the power of the system.

Exercise

1. Extend the facility above to compute $\int x^{n}e^{ax+b}\,dx$, where n is an integer and a and b are constants. You must handle the case $n = -1$ separately since

$$\int \frac{e^{x}}{x}\,dx = -\mathrm{Ei}(1,\,-x)\,.$$

Use the `ispoly` command from the Maple library to test for the expression $ax + b$ which is linear in x.

CHAPTER

Fundamentals

By now, you have no doubt written a number of procedures and found that Maple's programming language greatly extends the range of tasks you can tackle. Chapter 1 introduced a number of simple examples which you hopefully found intuitive and useful as models for creating many more of your own.

At some point or another, however, you may come across situations which seem curious. For instance, you may develop a sequence of commands which work reliably and correctly when you execute them interactively, but then no longer work when you incorporate them into a procedure by encapsulating them between a proc() and an end.

Even if this scenario does not sound familiar to you, you are likely to encounter it sooner or later if you write enough programs. Fortunately, the solution is almost always simple. A few fundamental rules dictate how Maple reads what you type. An understanding of these basic principles is particularly important within procedures, where you encounter different types of objects than what you may be familiar with.

Learning the basics is not difficult, especially if you understand five particularly important areas:

1. Maple's evaluation rules;

2. nested procedures, where Maple decides which variables are local, which are global, and which are parameters;

3. some particularly useful details of types: types which modify Maple's evaluation rules, structured types, and type matching;

4. data structures: understanding how to make effective use of them in order to best solve a problem;

5. and remember tables which, as you learned in chapter 1, can greatly increase the efficiency of your procedures.

In short, this chapter equips you with the fundamentals of Maple programming, thereby allowing you to understand and write non-trivial Maple code.

2.1 Evaluation Rules

Maple does not evaluate lines of code within procedures in quite the same way as it does if you enter those same lines in an interactive session. The rules for evaluation are simple, as this section demonstrates.

Of course, the evaluation rules within a procedure are different for good reasons, some which have to do with efficiency. In an interactive session, Maple evaluates most names and expressions completely. For instance, suppose that you assign a the value b and then assign b the value c. When you subsequently type a, Maple automatically follows your list of assignments to determine that the ultimate value of a is c.

```
> a := b;
```

$$a := b$$

```
> b := c;
```

$$b := c$$

```
> a + 1;
```

$$c + 1$$

In an interactive session, Maple tirelessly follows your chain of assignments, no matter how long the list. Within a procedure, however, Maple is sometimes not so diligent.

The substitution of assigned values for a name is called *evaluation* and each step in this process is known as an *evaluation level*. Using the `eval` command, you can explicitly ask Maple to perform evaluation of names to specific levels.

```
> eval(a, 1);
```

$$b$$

```
> eval(a, 2);
```

$$c$$

If you do not specify a number of levels, Maple evaluates the name to as many levels as exist.

```
> eval(a);
```

$$c$$

When you enter commands at the prompt, Maple usually evaluates the names as if you had enclosed each one in an `eval()`. The main exception is that evaluation stops whenever evaluating to one more level would turn the name into either a table, an array, or a procedure. The command a + 1 above is almost identical to `eval(a) + 1`.

In procedures, some rules are different. If you use the assignments above within a procedure, the result may surprise you.

```
> f := proc()
>        local a,b;
>        a := b;
>        b := c;
>        a + 1;
> end;
```

$$f := \mathbf{proc}() \ \mathbf{local} \ a, \ b; \ a := b; \ b := c; \ a + 1 \ \mathbf{end}$$

```
> f();
```

$$b + 1$$

The answer is b + 1 instead of c + 1, because a is a local variable and Maple evaluates local variables to only one level. The procedure behaves as if the final line were `eval(a,1) + 1`. Evaluating local variables fully is inefficient both in terms of time and memory. To evaluate a variable fully, Maple may have to follow a long list of assignments, resulting in a large expression.

The following sections introduce Maple's evaluation rules systematically. They discuss what types of variables can exist within a procedure and the evaluation rules applied to each.

Parameters

Chapter 1 introduces you to local and global variables, but procedures have a more fundamental type of variable: parameters. Parameters are variables whose name appears between the parentheses of a `proc()` statement. These have a special role within procedures, as Maple replaces them with arguments when you execute the procedure.

Examine the following procedure which squares its first argument and assigns the answer to the second argument, which must be a name.

```
> sqr1 := proc(x::anything, y::name)
>               y := x^2;
> end;
```

$$sqr1 := \textbf{proc}(x::anything,\ y::name)\ y := x^2\ \textbf{end}$$

```
> sqr1(d, ans);
```

$$d^2$$

```
> ans;
```

$$d^2$$

The procedure squares the value of d and assigns the result to the name ans. Try the procedure again, but this time use the name a which Maple earlier assigned the value b. Remember to reset ans to a name first.

```
> ans := 'ans';
```

$$ans := ans$$

```
> sqr1(a, ans);
```

$$c^2$$

```
> ans;
```

$$c^2$$

From the answer, Maple clearly remembers that you assigned b to the name a, and c to the name b. When did this evaluation occur?

To determine when, you must examine the value of x as soon as Maple enters the procedure. Use the debugger to get Maple to stop just after entering sqr1.

```
> stopat(sqr1);
```

$$[sqr1]$$

```
> ans := 'ans':
> sqr1(a, ans);

sqr1:
    1*   y := x^2
```

The value of the formal parameter x is c.

```
> x

c
sqr1:
```

```
    1*   y := x^2
> cont
```

$$c^2$$

```
> unstopat(sqr1):
```

In fact, Maple evaluates the arguments *before* invoking the procedure.

The steps Maple takes are best thought of in the following manner. When you call a procedure, Maple evaluates the arguments appropriately, given the context in which the call occurs. For example, if you call sqr1 from inside a procedure, then Maple evaluates a to one level. Thus, in the procedure g below, Maple evaluates a to b rather than to c.

```
> g := proc()
>        local a,b,ans;
>        a := b;
>        b := c;
>        sqr1(a,ans);
> end;
```

$$g :=$$
$$\textbf{proc}()\ \textbf{local}\ a,\ b,\ ans;\ a := b;\ b := c;\ \text{sqr1}(a,\ ans)\ \textbf{end}$$

```
> g();
```

$$b^2$$

Whether you call a procedure from the interactive level or from inside a procedure, Maple evaluates the arguments before invoking the procedure. Once Maple evaluates the arguments, it replaces all occurrences of the procedure's formal parameters with the actual arguments. Then Maple invokes the procedure.

Because Maple only evaluates parameters once, you cannot use them like local variables. The author of procedure cube, below, forgot that Maple does not re-evaluate parameters.

```
> cube := proc(x::anything, y::name)
>              y := x^3;
>              y;
> end:
```

When you call cube as below, Maple does assign ans the value 2^3, but the procedure returns the name ans rather than its value.

```
> ans := 'ans';;
```

$$ans := ans$$

```
> cube(2, ans);
```

$$ans$$

```
> ans;
```

$$8$$

Maple replaces each y with ans, but Maple does not evaluate these occurrences of ans again. Thus, the final line of cube returns the name ans, not the value that Maple assigned to ans.

Use parameters for two purposes: to pass information into the procedure and to return information from it. You may think of parameters as an objects evaluated to *zero* levels.

Local Variables

Local variables are temporary storage places within a procedure. You can create local variables using the `local` declaration statement at the beginning of a procedure. If you do not declare whether a variable is local or global, Maple decides for you. If you make an assignment to a variable within a procedure then Maple assumes that it should be `local`. A local variable is different from any other variable, whether global or local to another procedure, even if they have the same name. The rules for determining local variables become a little more involved when *nested procedures* are written, but the basic concepts are similar. See *Nested Procedures* on page 49 for more details.

Maple only evaluates local variables to one level.

```
> f := proc()
>        local a,b;
>        a := b;
>        b := c;
>        a + 1;
> end;
```

$$f := \mathbf{proc}() \ \mathbf{local} \ a, \ b; \ a := b; \ b := c; \ a + 1 \ \mathbf{end}$$

When you invoke f, Maple evaluates the a in a+1 one level to b.

```
> f();
```

$$b + 1$$

Maple always uses last-name evaluation for tables, arrays, and procedures. Therefore, if you assign a local variable to a table, an array, or a procedure, Maple does not evaluate that variable unless you use `eval`. Maple creates the local variables of a procedure each time you call the procedure. Thus, local variables are local to a specific invocation of a procedure.

If you have not written many programs you might think that one level evaluation of local variables is a serious limitation, but in fact code which requires further evaluation of local variables is difficult to understand, and is unnecessary. Moreover, because Maple does not attempt further evaluations, it saves many steps, causing procedures to run faster.

Global Variables

Global variables are available from inside any procedure in Maple as well as at the interactive level. Indeed, any name you use at the interactive level is a global variable, allowing you to write a procedure which assigns a value to a variable that is accessible again later from within another procedure, the same procedure, or at the interactive level.

```
> h := proc()
>       global x;
>       x := 5;
> end:
> h();
```

$$5$$

```
> x;
```

$$5$$

Within procedures, use global variables with caution. The procedure h assigns a value to the global variable x but it does not leave any warning in your worksheet. If you then use x thinking that it is an unknown, you can get bewildering error messages.

```
> diff( x^2, x);
```

```
Error,
wrong number (or type) of parameters in function diff
```

Moreover, if you write yet another procedure which uses the global variable x, then the two procedures may use the same x in incompatible ways.

Whether within a procedure or at the interactive level, Maple always applies the same evaluation rules to global variables. It evaluates all global names fully, except when the value of such a variable is a table, an array, or a procedure, in which case, Maple halts its evaluation at the last name in the chain of assignments. This evaluation rule is called *last-name evaluation*.

Hence, *Maple evaluates parameters to zero levels, local variables to one level, and global variables fully*, except for last-name evaluation.

As with local variables, the rules for determining which variables are global are fully described in *Nested Procedures* on page 49.

Exceptions

This section describes two exceptions of particular note to the rules for evaluation.

The Ditto Operator The *ditto operator*, %, which recalls the last result, is local to procedures but Maple evaluates it *fully*. When you invoke a procedure, Maple initializes the local version of % to NULL.

```
> f := proc()
>    local a,b;
>    print( "Initially [%] has the value", [%] );
>    a := b;
>    b := c;
>    a + 1;
>    print( "Now [%] has the value", [%] );
> end:
> f();
```

$$\text{“Initially [\%] has the value”}, \ []$$
$$\text{“Now [\%] has the value”}, \ [c+1]$$

The same special rules apply to the %% and %%% operators. Using local variables instead of ditto operators makes your procedures easier to read and debug.

Environment Variables The variable Digits, which determines the number of digits that Maple uses when calculating with floating-point numbers, is an example of an *environment variable*. Maple evaluates environment variables in the same manner it evaluates global variables; that is, Maple evaluates environment variables fully except for last-name evaluation. When a procedure returns, Maple resets all environment variables to the values they had when you invoked the procedure.

```
> f := proc()
>    print( "Entering f.  Digits is", Digits );
>    Digits := Digits + 13;
>    print( "Adding 13 to Digits yields", Digits );
> end:
> g := proc()
>    print( "Entering g.  Digits is", Digits );
>    Digits := 77;
>    print( "Calling f from g.  Digits is", Digits );
```

```
>    f();
>    print( "Back in g from f.  Digits is", Digits );
> end:
```

The default value of `Digits` is 10.

```
> Digits;
```

$$10$$

```
> g();
```

"Entering g. Digits is", 10

"Calling f from g. Digits is", 77

"Entering f. Digits is", 77

"Adding 13 to Digits yields", 90

"Back in g from f. Digits is", 77

When returning from g, Maple resets `Digits` to 10.

```
> Digits;
```

$$10$$

See `?environment` for a list of environment variables. You can also make your own environment variables: Maple considers any variable whose name begins with the four characters _Env to be an environment variable.

2.2 Nested Procedures

You can define a Maple procedure inside another Maple procedure. Indeed, you may commonly write such procedures without realizing you are writing nested procedures. In interactive sessions, you are no doubt familiar with using the `map` command to apply some operation to the elements of some type of structure. For example, you may want to divide each element of a list by a number, such as 8.

```
> lst := [8, 4, 2, 16]:
> map( x->x/8, lst);
```

$$\left[1, \frac{1}{2}, \frac{1}{4}, 2 \right]$$

The `map` command is also very useful inside a procedure. Consider another variation on this command which appears in the following procedure. The intent of this new procedure is to divide each element of a list by the first element of that list.

```
> nest := proc(x::list)
>    local v;
>    v := x[1];
>    map( y -> y/v, x );
> end:
```

```
> nest(lst);
```

$$\left[1, \frac{1}{2}, \frac{1}{4}, 2 \right]$$

Maple considers this use of map as an example of nested procedures and applies its lexical scoping rules, which declare the v within the call to map as the same v as in the outer procedure, nest.

The following section explains Maple's scoping rules. You will learn how Maple decides which variables are local to a procedure and which are global. Understanding Maple's evaluation rules for parameters, and for local and global variables, allows you to make full use of the Maple language.

Local or Global?

Usually when you write a procedure, you should explicitly declare which variables are global and which are local. Declaring the scope of the variables makes your procedure easier to read and debug. However, sometimes declaring the variables is not the way to go. In the nest procedure above, the variable in the map command gets its meaning from the surrounding procedure. What happens if you define this variable, v, as local to the invocation of the procedure within map?

```
> nest2 := proc(x::list)
>    local v;
>    v := x[1];
>    map( proc(y) local v; y/v; end, x );
> end:
> nest2(lst);
```

$$\left[\frac{8}{v}, \frac{4}{v}, \frac{2}{v}, \frac{16}{v} \right]$$

If you examine nest2 closely, you should be able to determine why it didn't work the same as nest. It is obvious that you don't want to have the variables declared at all within the inner procedure, so that it can get it's proper meaning from the enclosing procedure.

Only two possibilities exist: either a variable is local to a procedure and certain procedures that are completely within it, or it is global to the entire Maple session.

The method Maple uses for determining whether a variable is local or global can be summarized as: The name of the variable is searched for among the parameters, `local` declarations, and `global` declarations of the procedure, and then among the parameters, `local` and `global` declarations, and implicitly declared local variables of any surrounding procedure(s), from the inside out. If found, that specifies the binding of the variable.

If, using the above rule, Maple cannot determine whether a variables should be global or local, the following default decisions are made for you. *If a variable appears on the left-hand side of an explicit assignment or as the controlling variable of a* `for` *loop, then Maple assumes that you intend the variable to be local.* Otherwise, Maple assumes that the variable is global to the whole session. In particular, Maple assumes by default that the variables you only pass as arguments to other procedures, which may set their values, are global.

The Quick-Sort Algorithm

Sorting algorithms is one of the principal types of routines of interest to computer scientists. Even if you have never formally studied them you can appreciate that many things need sorting. Sorting a few numbers is quick and easy no matter what approach you use, but sorting large amounts of data can be very time consuming; thus, finding efficient methods is important.

The quick-sort algorithm below is a classic algorithm. The key to understanding this algorithm is to understand the operation of partitioning. This involves choosing any one number from the array that you are about to sort. Then, you shove the numbers in the array that are less than the number that you chose to one end of the array, and numbers greater to the other end. Lastly, you insert the chosen number between these two groups.

At the end of the partitioning, you have not yet entirely sorted the array, because the numbers less or greater than the one you chose may still be in their original order. This procedure divides the array into two smaller arrays which are easier to sort than the original larger one. The partitioning operation has thus made the work of sorting much easier. Better yet, you can bring the array one step closer in the sorting process by partitioning each of the two smaller arrays. This operation produces four smaller arrays. You sort the entire array by repeatedly partitioning the smaller arrays.

The partition procedure uses an array to store the list because you can change the elements of an array directly. Thus, you can sort the array in place and not waste any space generating extra copies.

The quicksort procedure is easier to understand if you look at the procedure partition in isolation first. This procedure accepts an array of numbers and two integers. The two integers are element numbers of the array, indicating the portion of the array to partition. While you could possibly choose any of the numbers in the array to partition around, this procedure chooses the last element of the section of the array for that purpose, namely A[n]. The intentional omission of global and local statements is to show which variables Maple thinks are local and which global by default.

```
> partition := proc(A::array(1, numeric),
>                    m::integer, n::integer)
>     i := m;
>     j := n;
>     x := A[j];
>     while i<j do
>        if A[i]>x then
>           A[j] := A[i];
>           j := j-1;
>           A[i] := A[j];
>        else
>           i := i+1;
>        fi;
>     od;
>     A[j] := x;
>     eval(A);
> end:

Warning, 'i' is implicitly declared local
Warning, 'j' is implicitly declared local
Warning, 'x' is implicitly declared local
```

Maple declares i, j, and x local because the partition procedure contains explicit assignments to those variables. partition also assigns explicitly to A, but A is a parameter, not a local variable. Because you do not assign to the name eval, Maple makes it the global name which refers to the eval command.

After partitioning the array a below, all the elements less than 3 precede 3 but they are in no particular order; similarly, the elements larger than 3 come after 3.

```
> a := array( [2,4,1,5,3] );
```

$$a := [2, 4, 1, 5, 3]$$

```
> partition( a, 1, 5);
```

$$[2, 1, 3, 5, 4]$$

The partition procedure modifies its first argument, thus changing a.

```
> eval(a);
```

$$[2, 1, 3, 5, 4]$$

The final step in assembling the quick-sort procedure is to insert the partition procedure within an outer procedure. The outer procedure first defines the partition subprocedure, then partitions the array. Ordinarily, you might wish to avoid inserting one procedure within another. However, you will encounter situations in chapter 3, where you will find it necessary to do so. Since the next step is to partition each of the two subarrays by calling quicksort recursively, partition must return the location of the element which divides the partition.

```
> quicksort := proc(A::array(1, numeric),
>                   m::integer, n::integer)
>    local partition, p;
>
>    partition := proc(m,n)
>       i := m;
>       j := n;
>       x := A[j];
>       while i<j do
>          if A[i]>x then
>             A[j] := A[i];
>             j := j-1;
>             A[i] := A[j];
>          else
>             i := i+1;
>          fi;
>       od;
>       A[j] := x;
>       p := j;
>    end:
>
>    if m<n then    # if m>=n there is nothing to do
>       partition(m, n);
>       quicksort(A, m, p-1);
>       quicksort(A, p+1, n);
>    fi;
>    eval(A);
> end:
Warning, 'i' is implicitly declared local
```

```
Warning, 'j' is implicitly declared local
Warning, 'x' is implicitly declared local

> a := array( [2,4,1,5,3] );
```

$$a := [2, 4, 1, 5, 3]$$

```
> quicksort( a, 1, 5);
```

$$[1, 2, 3, 4, 5]$$

```
> eval(a);
```

$$[1, 2, 3, 4, 5]$$

Maple determines that the A and p variables in the partition subprocedure are defined by the parameter and local variable (respectively) from the outer quicksort procedure and everything works as planned. We could also have passed A as a parameter to the partition subprocedure (as we did when partition was a stand-alone procedure), but because of the scoping rules, it wasn't necessary to do so.

Creating a Uniform Random Number Generator

If you want to use Maple to simulate physical experiments, you likely need a random number generator. The uniform distribution is particularly simple: any real number in a given range is equally likely. Thus, a *uniform random number generator* is a procedure that returns a random floating-point number within a certain range. This section develops the procedure, uniform, which creates uniform random number generators.

The rand command generates a procedure which returns random *integers*. For example, rand(4..7) generates a procedure that returns random integers between 4 and 7, inclusive.

```
> f := rand(4..7):
> seq( f(), i=1..20 );
```

$$5, 6, 5, 7, 4, 6, 5, 4, 5, 5, 7, 7, 5, 4, 6, 5, 4, 5, 7, 5$$

The uniform procedure should be similar to rand but should return floating-point numbers rather than integers. You can use rand to generate random floating-point numbers between 4 and 7 by multiplying and dividing by 10^Digits.

```
> f := rand( 4*10^Digits..7*10^Digits ) / 10^Digits:
> f();
```

$$\frac{12210706011}{2000000000}$$

The procedure f returns fractions rather than floating-point numbers so you must compose it with evalf; that is, use evalf(f()). You can perform this operation better using Maple's composition operator, @.

```
> (evalf @ f)();
```

$$6.648630719$$

The uniform procedure below uses evalf to evaluate the constants in the range specification, r, to floating-point numbers, the map command to multiply both endpoints of the range by 10^Digits, and round to round the results to integers.

```
> uniform := proc( r::constant..constant )
>    local intrange, f;
>    intrange := map( x -> round(x*10^Digits), evalf(r) );
>    f := rand( intrange );
>    (evalf @ eval(f)) / 10^Digits;
> end:
```

You can now generate random floating-point numbers between 4 and 7.

```
> U := uniform(4..7):
> seq( U(), i=1..20 );
```

$$4.559076346, 4.939267370, 5.542851096, 4.260060897,$$
$$4.976009937, 5.598293374, 4.547350945,$$
$$5.647078832, 5.133877918, 5.249590037,$$
$$4.120953928, 6.836344299, 5.374608653,$$
$$4.586266491, 5.481365622, 5.384244382,$$
$$5.190575456, 5.207535837, 5.553710879,$$
$$4.163815544$$

The uniform procedure suffers from a serious flaw: uniform uses the current value of Digits to construct intrange; thus, U depends on the value of Digits when uniform creates it. On the other hand, the evalf command within U uses the value of Digits that is current when you invoke U. These two values are not always identical. The proper design choice here is that U should depend only on the value of Digits when you invoke U. The version of uniform below accomplishes this by placing all the computation inside the procedure that uniform returns.

```
> uniform := proc( r::constant..constant )
>
```

```
>    proc()
>       local intrange, f;
>       intrange := map( x -> round(x*10^Digits),
>                           evalf(r) );
>       f := rand( intrange );
>       evalf( f()/10^Digits );
>    end;
> end:
```

The r within the inner proc is not declared as local or global, so it becomes the same r as the parameter to the outer proc.

The procedure that uniform generates is now independent of the value of Digits at the time you invoke uniform.

```
> U := uniform( cos(2)..sin(1) ):
> Digits := 5:
> seq( U(), i=1..8 );
```

$$-.17503, \ -.11221, \ -.15794, \ -.18007, \ .38662, \ -.40436,$$

$$.094310, \ .17760$$

This section introduced you to the rules Maple uses to decide which variables are global or local. You have also seen the principal implications of these rules. In particular, it introduced you to the tools available for writing nested procedures.

2.3 Types

Types that Modify Evaluation Rules

Evaluation Rules on page 42 introduces the details of how Maple evaluates different kinds of variables within a procedure: Maple evaluates global variables fully (except for last-name evaluation) and local variables one level. Maple evaluates the arguments to a procedure, depending upon the circumstances, *before* invoking the procedure, and then simply substitutes the actual parameters for the formal parameters within the procedure without any further evaluation. All these rules seem to imply that nothing within the procedure in any way affects the evaluation of arguments which occurs *before* Maple invokes the procedure. In reality, the exceptions provide convenient methods for controlling the evaluation of arguments which make your procedures' behavior more intuitive to you. They also prevent evaluation which would result in the loss of information you wish available within your procedure.

Maple uses different evaluation rules for some of its own commands, for example, the `evaln` command. You have no doubt used this command to clear the value of previously defined variables. If this command were to evaluate its argument normally, it would be of no use for this purpose. For example, if you assign x the value π, then Maple evaluates x to π whenever you use the variable x.

```
> x := Pi;
```

$$x := \pi$$

```
> cos(x);
```

$$-1$$

If Maple behaved the same way when you type `evaln(x)`, then Maple would pass the value π to `evaln`, losing all references to the name x. Therefore Maple evaluates the argument to `evaln` in a special way: it evaluates the argument to a name, not to the value that name may have.

```
> x := evaln(x);
```

$$x := x$$

```
> cos(x);
```

$$\cos(x)$$

You will find it useful to write your own procedures which exhibit this behavior. You may want to write a procedure which returns a value by assigning it to one of the arguments. *Parameters* on page 43 describes such a procedure, `sqr1`, but each time you call `sqr1` you must take care to pass it an unassigned name.

```
> sqr1:= proc(x::anything, y::name)
>    y := x^2;
> end:
```

This procedure works fine the first time you call it. However, you must make sure that the second argument is indeed a name; otherwise, an error results. In the example below, the error occurs because, upon the second attempt, `ans` has the value 9.

```
> ans;
```

$$ans$$

```
> sqr1(3, ans);
```

$$9$$

```
> ans;
```

$$9$$

```
> sqr1(4, ans);
```

```
Error, sqr1 expects its 2nd argument, y,
to be of type name, but received 9
```

You have two ways around this problem. The first is to use either single quotes or the evaln command to ensure that Maple passes a name and not a value. The second is to declare the parameter to be of type evaln.

Just like the evaln command, declaring a parameter to be of type evaln causes Maple to evaluate that argument to a name, so you do not have to worry about evaluation when you use the procedure.

```
> cube := proc(x::anything, y::evaln)
>    y := x^3;
> end:
> ans;
```

$$9$$

```
> cube(5, ans);
```

$$125$$

```
> ans;
```

$$125$$

In the above case, Maple passes the name ans to the cube procedure instead of the value 9.

Using the evaln declaration is generally a good idea. It ensures that your procedures do what you expect instead of returning cryptic error messages. However, some Maple programmers like to use the single quotes. When the call to the procedure is within a procedure itself, the presence of the single quotes is a reminder that you are assigning a value to a parameter. However, if you plan to use your procedure interactively, you will find using evaln far more convenient.

A second type which modifies Maple's evaluation rules is uneval. Where evaln makes Maple evaluate the argument to a name, uneval leaves the argument unevaluated. This type is useful for two reasons. First, sometimes you wish to write a procedure which treats a structure as an object and does not require knowledge of the details. Second, sometimes expanding the argument within the procedure is useful. You may want to write a version of the map command which is capable of mapping over sequences. The standard map command built into Maple is not capable of this because it evaluates its second argument. If the second argument is the name of a sequence, Maple evaluates the name to the sequence before invoking map. Since Maple flattens sequences of sequences, it passes only the first element of the sequence as the second argument to map and the other elements become additional arguments.

The smap procedure below uses an uneval declaration to tell Maple not to evaluate its second argument. Once inside the procedure, the eval command fully evaluates S. The whattype command returns exprseq if you pass it a sequence.

```
> whattype( a, b, c );
```

$$exprseq$$

If S is not a sequence, smap simply calls map. args[3..-1] is the sequence of arguments to smap after S. If S is a sequence, enclosing it in square brackets forms a list. You can then map f onto the list and use the selection operator, [], to turn the resulting list back into a sequence.

```
> smap := proc( f::anything, S::uneval )
>    local s;
>    s := eval(S);
>    if whattype(s) = 'exprseq' then
>        map( f, [s], args[3..-1] )[];
>    else
>        map( f, s, args[3..-1] );
>    fi;
> end:
```

Now you can map over sequences as well as lists, sets, and other expressions.

```
> S := 1,2,3,4;
```

$$S := 1, 2, 3, 4$$

```
> smap(f, S, x, y);
```

$$f(1, x, y), f(2, x, y), f(3, x, y), f(4, x, y)$$

```
> smap(f, [a,b,c], x, y);
```

$$[f(a, x, y), f(b, x, y), f(c, x, y)]$$

Both evaln and uneval greatly extend the flexibility of Maple's programming language and the types of procedures you can write.

Structured Types

Sometimes a simple type check, either through declared formal parameters or explicitly with the type command, does not provide enough information. A simple check tells you that 2^x is an exponentiation but it does not distinguish between 2^x and x^2.

```
> type( 2^x, '^' );
```

$$true$$

```
> type( x^2, '^' );
```

$$true$$

To make such distinctions you need *structured types*. For example, 2 is a constant and x is a name, so 2^x has type constant^name but x^2 does not.

```
> type( 2^x, constant^name );
```

$$true$$

```
> type( x^2, constant^name );
```

$$false$$

Suppose you want to solve a set of equations. Before proceeding you want to remove any equations that are trivially true, like $4 = 4$. Thus, you need to write a procedure that accepts a set of equations as input. The procedure nontrivial below uses automatic type checking to ensure that the argument is indeed a set of equations.

```
> nontrivial := proc( S::set( '=' ) )
>     remove( evalb, S );
> end:
> nontrivial( { x^2+2*x+1=0, y=y, z=2/x } );
```

$$\left\{ x^2 + 2x + 1 = 0, \ z = \frac{2}{x} \right\}$$

You can easily extend nontrivial so that it accepts general relations rather than just equations, and so that it allows both sets and lists of relations. An expression matches a set of types if it matches one of the types in the set.

```
> nontrivial := proc( S::{ set(relation), list(relation) } )
>     remove( evalb, S );
> end:
> nontrivial( [ 2<=78, 1/x=9 ] );
```

$$\left[\frac{1}{x} = 9 \right]$$

You can extend `nontrivial` even further: if an element in S is not a relation but an algebraic expression, f, then `nontrivial` should treat it as the equation $f = 0$.

```
> nontrivial := proc( S::{  set( {relation, algebraic} ),
>                           list( {relation, algebraic} ) } )
>    local istrivial;
>    istrivial := proc(x)
>        if type(x, relation) then evalb(x);
>        else evalb( x=0 );
>        fi;
>    end;
>    remove( istrivial, S );
> end:
> nontrivial( [ x^2+2*x+1, 23>2, x=-1, y-y ] );
```
$$[x^2 + 2x + 1,\ x = -1]$$

Automatic type checking is a very powerful tool. It allows you to do a large amount of checking for invalid arguments automatically. You should make using it a habit. Structured types allow checking even when you design a procedure to accept a variety of inputs, or to rely on a particular structure in its arguments.

Automatic type checking has two weaknesses. First, if the structure of the type is complicated, permitting several structures, then the code for the type checking can become cumbersome. The second is that Maple does not save any of the information about the structure of the arguments. It parses and checks them, but then the structure is lost. If you wish to extract a particular component of the structure you must write more code to do so.

The complexity of the types is rarely of concern in practice. A procedure which relys on arguments with a complicated structure is usually hard to use. The `typematch` command addresses the duplication of effort in parsing the arguments. This command provides a more flexible alternative method of type checking.

Type Matching

Integration by Parts on page 35 describes the following pair of procedures that implement indefinite integration of any polynomial times e^x.

```
> IntExpMonomial := proc(n::nonnegint, x::name)
>    if n=0 then RETURN( exp(x) ) fi;
>    x^n*exp(x) - n*IntExpMonomial(n-1, x);
> end:
> IntExpPolynomial := proc(p::polynom, x::name)
>    local i, result;
```

```
>    result := add( coeff(p, x, i)*IntExpMonomial(i, x),
>                    i=0..degree(p, x) );
>    collect(result, exp(x));
> end:
```

You may want to modify `IntExpPolynomial` so that it can also perform definite integration. The new version of `IntExpPolynomial` should allow its second argument to be a name, in which case `IntExpPolynomial` should perform indefinite integration, or the form *name=range*. You could use the **type** command and **if** statements to do this, but then the procedure becomes difficult to read.

```
> IntExpPolynomial := proc(p::polynom, xx::{name, name=range})
>    local i, result, x, a, b;
>    if type(xx, name) then
>        x:=xx;
>    else
>        x  := lhs(xx);
>        a  := lhs(rhs(xx));
>        b  := rhs(rhs(xx));
>    fi;
>    result := add( coeff(p, x, i)*IntExpMonomial(i, x),
>                    i=0..degree(p, x) );
>    if type(xx, name) then
>        collect(result, exp(x));
>    else
>        eval(result, x=b) - eval(result, x=a);
>    fi;
> end:
```

Using the `typematch` command makes your procedure much easier to read. The `typematch` command not only tests if an expression matches a certain type, it can also assign variables to pieces of the expression. Below, `typematch` checks that expr is of the form name=integer..integer *and* it assigns the name to y, the left-hand limit to a, and the right-hand limit to b.

```
> expr := myvar=1..6;
```

$$expr := myvar = 1..6$$

```
> typematch( expr, y::name=a::integer..b::integer );
```

$$true$$

```
> y, a, b;
```

$$myvar, 1, 6$$

The version of `IntExpPolynomial` below uses the `typematch` command.

```
> IntExpPolynomial := proc(p::polynom, expr::anything )
>     local i, result, x, a, b;
>     if not typematch( expr, {x::name,
>             x::name=a::anything..b::anything} ) then
>        ERROR( "expects a name or name=range but received",
>                   expr );
>     fi;
>     result := add( coeff(p, x, i)*IntExpMonomial(i, x),
>                    i=0..degree(p, x) );
>     if type(expr, name) then
>        collect(result, exp(x));
>     else
>        eval(result, x=b) - eval(result, x=a);
>     fi;
> end:
```

Now `IntExpPolynomial` can perform definite, as well as indefinite, integrals.

```
> IntExpPolynomial( x^2+x^5*(1-x), x=1..2 );
```

$$-118\,e^2 + 308\,e$$

```
> IntExpPolynomial( x^2*(x-1), x);
```

$$(-4\,x^2 + 8\,x - 8 + x^3)\,e^x$$

2.4 Choosing a Data Structure: Connected Graphs

When writing programs you have to decide how to represent the data. Sometimes the choice is straightforward but often it requires considerable thought and planning. Some choices of data structure may make your procedures more efficient or easier to write and debug. No doubt you are familiar with Maple's many available data structures, such as sequences, lists, tables, and sets.

This section introduces you to a variety of structures and their advantages. This section also illustrates by means of an example the problem of choosing a data structure. Suppose you have a number of cities with roads between them. Write a procedure which determines whether you can travel between any two cities.

You can express this problem in terms of graph theory, and Maple has a `networks` package that helps you work with graphs and more general structures. You do not need to understand graph theory or the `networks` package to benefit from the examples in this section; these examples primarily use the networks package as a shortcut to the drawing of *G*, below.

```
> with(networks):
```

Make a new graph *G* and add a few cities, or *vertices*, in the terminology of graph theory.

```
> new(G):
> cities := {Zurich, Rome, Paris, Berlin, Vienna};
```

$$cities := \{Zurich, \ Rome, \ Paris, \ Vienna, \ Berlin\}$$

```
> addvertex(cities, G);
```

$$Zurich, \ Rome, \ Paris, \ Vienna, \ Berlin$$

Add roads between Zurich and each of Paris, Berlin, and Vienna. The connect command names the roads *e*1, *e*2, and *e*3.

```
> connect( {Zurich}, {Paris, Berlin, Vienna}, G );
```

$$e1, \ e2, \ e3$$

Add roads between Rome and Zurich and between Berlin and both Paris and Vienna.

```
> connect( {Rome}, {Zurich}, G);
```

$$e4$$

```
> connect( {Berlin}, {Vienna, Paris}, G);
```

$$e5, \ e6$$

Now draw the graph *G*.

```
> draw(G);
```

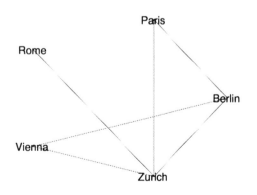

If you look at the drawing above, you can convince yourself that, in this particular case, you could travel between any two cities. Instead of visual inspection, you can also use the `connectivity` command.

```
> evalb( connectivity(G) > 0 );
```

true

The data structures that the `networks` package uses are quite involved, because that package supports more general structures than you need in this example. The question then is: how would *you* represent the cities and roads? Since cities have distinct names and the order of the cities is irrelevant, you could represent the cities as a set of names.

```
> vertices(G);
```

{Zurich, Rome, Paris, Vienna, Berlin}

The `networks` package assigns distinct names to the roads, so it can also represent them as set of names.

```
> edges(G);
```

{e1, e2, e3, e4, e5, e6}

You can also represent a road as the set consisting of the two cities the road connects.

```
> ends(e2, G);
```

{Zurich, Vienna}

Thus, you can represent the roads as a set of sets.

```
> roads := map( ends, edges(G), G);
```

roads := {{*Zurich, Rome*}, {*Zurich, Paris*},

{*Zurich, Vienna*}, {*Zurich, Berlin*}, {*Paris, Berlin*},

{*Vienna, Berlin*}}

Unfortunately, if you want to know which cities are directly connected to Rome, for example, you have to search through the whole set of roads. Therefore, representing the data as a set of cities and a set of roads is computationally inefficient for determining whether you can travel between any two cities.

You can also represent the data as an *adjacency matrix*: a square matrix with a row for each city. The (i, j)th entry in the matrix is 1 if the ith and the jth city have a road between them, and 0 otherwise. The following is the adjacency matrix for the graph G.

```
> adjacency(G);
```

$$\begin{bmatrix} 0 & 1 & 0 & 1 & 1 \\ 1 & 0 & 0 & 0 & 1 \\ 0 & 0 & 0 & 0 & 1 \\ 1 & 0 & 0 & 0 & 1 \\ 1 & 1 & 1 & 1 & 0 \end{bmatrix}$$

The adjacency matrix is an inefficient representation if few roads exist relative to the number of cities. In that case, the matrix contains many zeros, representing an overall lack of roads. Also, though each row in the matrix corresponds to a city, you cannot tell which row corresponds to which city.

Here is yet another way of representing the cities and roads: Paris has two roads between it and both Zurich and Berlin; thus, Berlin and Zurich are the neighbors of Paris.

```
> neighbors(Paris, G);
```

$$\{Zurich, Berlin\}$$

You can represent the data as a table of neighbors; one entry should be in the table for each city.

```
> T := table( map( v -> (v)=neighbors(v,G), cities ) );
```

$$T := \text{table}([$$

$$Zurich = \{Rome, Paris, Vienna, Berlin\}$$

$$Rome = \{Zurich\}$$

$$Paris = \{Zurich, Berlin\}$$

$$Vienna = \{Zurich, Berlin\}$$

$$Berlin = \{Zurich, Paris, Vienna\}$$

$$])$$

The representation of a system of cities and roads as a table of neighbors is ideally suited to answering the question of whether it is possible to travel between any two cities. You can begin at one city. The table allows you to efficiently find the neighboring cities to which you can travel. Similarly, you can find the neighbors of the neighbors, and thus you can quickly determine how far you can travel.

The connected procedure below determines whether you can travel between any two cities. It uses the indices command to extract the set of cities from the table.

```
> indices(T);
```

$$[Zurich], [Rome], [Paris], [Vienna], [Berlin]$$

Since the `indices` command returns a sequence of lists, you must use the `op` and `map` command to generate the set.

```
> map( op, {%} );
```

$$\{Zurich,\ Rome,\ Paris,\ \ Vienna,\ Berlin\}$$

The `connected` procedure initially visits the first city, v. Then `connected` adds v to the set of cities that it has already visited and v's neighbors to the set of cities to which it can travel. As long as `connected` can travel to more cities, it does so. When `connected` has no more new cities to which it can travel, it determines whether it has seen all the cities.

```
> connected := proc( T::table )
>    local canvisit, seen, v, V;
>    V := map( op, { indices(T) } );
>    seen := {};
>    canvisit := { V[1] };
>    while canvisit <> {} do
>        v := canvisit[1];
>        seen := seen union {v};
>        canvisit := ( canvisit union T[v] ) minus seen;
>    od;
>    evalb( seen = V );
> end:
> connected(T);
```

$$true$$

You can add the cities Montreal, Toronto, and Waterloo, and the highway between them.

```
> T[Waterloo] := {Toronto};
```

$$T_{Waterloo} := \{Toronto\}$$

```
> T[Toronto] := {Waterloo, Montreal};
```

$$T_{Toronto} := \{Montreal,\ Waterloo\ \}$$

```
> T[Montreal] := {Toronto};
```

$$T_{Montreal} := \{Toronto\}$$

Now you can no longer travel between any two cities; for example, you cannot travel from Paris to Waterloo.

```
> connected(T);
```

false

Exercises

1. The system of cities and roads above splits naturally into two components: the Canadian cities and the roads between them, and the European cities and the roads between them. In each component you can travel between any two cities but you cannot travel between the two components. Write a procedure that, given a table of neighbors, splits the system into such components. You may want to think about the form in which the procedure should return its result.

2. The `connected` procedure above cannot handle the empty table of neighbors.

```
> connected( table() );

Error, (in connected) invalid subscript selector
```

Correct this shortcoming.

The importance of this example is not to teach you about networks, but to emphasize how the choice of data structures suited to the problem allows you to create an efficient and concise version of the procedure `connected`. Sets and tables were the best choices here. The best choice for a problem that you wish to tackle may be very different. Before writing code to perform your task, pause to consider which structures best suit your needs. A good program design begins with choosing structures and methods which mirror the data and task at hand.

2.5 Remember Tables

Sometimes procedures are designed such that they are called repeatedly with the same arguments. Each time, Maple has to recompute the same answer, unless you take advantage of Maple's concept of *remember tables*.

Any Maple procedure can have a remember table. The purpose of a remember table is to improve the efficiency of a procedure by storing previous results so that Maple can retrieve them from the table instead of recomputing them.

A remember table uses the sequence of actual parameters to the procedure call as the table index, and the results of the procedure calls as the table values. Whenever you invoke a procedure which has a remember table, Maple searches the table for an index which is the sequence of actual parameters. If such an index is found, it returns the corresponding value in the table as the result of the procedure call. Otherwise, Maple executes the body of the procedure.

Maple tables are hash tables, so looking up previously computed results is very fast. The purpose of remember tables is to make use of fast table lookup in order to avoid recomputing results. Since remember tables can become large, they are most useful when procedures need the same results repeatedly and the results are expensive to compute.

The remember Option

Use the remember option to indicate to Maple that it should store the result of a call to a procedure in a remember table. The Fibonacci procedure in *The* RETURN *Command* on page 22 is an example of a recursive procedure with the remember option.

```
> Fibonacci := proc(n::nonnegint)
>     option remember;
>     if n<2 then RETURN(n) fi;
>     Fibonacci(n-1) + Fibonacci(n-2);
> end:
```

Recursive Procedures on page 20 demonstrates that the Fibonacci procedure is very slow without the remember option, since it must compute the lower Fibonacci numbers many times.

When you ask Fibonacci to calculate the third Fibonacci number, it adds four entries to its remember table. The remember table is the fourth operand of a procedure.

```
> Fibonacci(3);
```

$$2$$

```
> op(4, eval(Fibonacci));
```

$$table([$$
$$3 = 2$$
$$0 = 0$$
$$1 = 1$$
$$2 = 1$$
$$])$$

Adding Entries Explicitly

You can also define entries in procedure remember tables yourself. To do so, use the following syntax.

$$f(x) := result:$$

Below is another procedure which generates the Fibonacci numbers. The fib procedure uses two entries in its remember table, where Fibonacci uses an if statement.

```
> fib := proc(n::nonnegint)
>    option remember;
>    fib(n-1) + fib(n-2);
> end:
> fib(0) := 0:
> fib(1) := 1:
```

You must add entries in the remember table *after* making the procedure. The option remember statement does *not* create the remember table, but rather asks Maple to automatically *add* entries to it. The procedure works without this option, but less efficiently.

You could even write a procedure which chooses which values to add to its remember table. The following version of fib only adds entries to its remember table when you call it with an odd-valued argument.

```
> fib := proc(n::nonnegint)
>    if type(n,odd) then
>       fib(n) := fib(n-1) + fib(n-2);
>    else
>       fib(n-1) + fib(n-2);
>    fi;
> end:
> fib(0) := 0:
> fib(1) := 1:
> fib(9);
```

$$34$$

```
> op(4, eval(fib));
```

$$table([$$
$$3 = 2$$
$$7 = 13$$
$$0 = 0$$

$$1 = 1$$
$$5 = 5$$
$$9 = 34$$
$$])$$

As in this case, sometimes you can dramatically improve the efficiency of a procedure by remembering only some of the values instead of none.

Removing Entries from a Remember Table

You can remove entries from a remember table in the same manner you remove entries from any other table: assign a table entry to its own name. The evaln command evaluates an object to its name.

```
> T := op(4, eval(fib) );
```

$$T := \text{table}([$$
$$3 = 2$$
$$7 = 13$$
$$0 = 0$$
$$1 = 1$$
$$5 = 5$$
$$9 = 34$$
$$])$$

```
> T[7] := evaln( T[7] );
```

$$T_7 := T_7$$

Now the fib procedure's remember table has only five entries.

```
> op(4, eval(fib) );
```

$$\text{table}([$$
$$3 = 2$$
$$0 = 0$$
$$1 = 1$$
$$5 = 5$$

$$9 = 34$$

$$])$$

Maple can also remove remember table entries automatically. If you give your procedure the `system` option, then Maple may remove entries in the procedure's remember table when Maple performs a garbage collection. Thus, you should *never* give the `system` option to procedures like `fib` that rely on entries in its remember table to terminate.

You can remove a procedure's remember table altogether by substituting NULL for the procedure's fourth operand.

```
> subsop( 4=NULL, eval(Fibonacci) ):
> op(4, eval(Fibonacci));
```

You should only use remember tables with procedures whose results depend exclusively on parameters. The procedure below depends on the value of the environment variable `Digits`.

```
> f := proc(x::constant)
>    option remember;
>    evalf(x);
> end:
> f(Pi);
```

$$3.141592654$$

Even if you change the value of `Digits`, `f(Pi)` remains unchanged because Maple retrieves the value from the remember table.

```
> Digits := Digits + 34;
```

$$Digits := 44$$

```
> f(Pi);
```

$$3.141592654$$

2.6 Conclusion

A thorough understanding of the concepts in this chapter will provide you with an excellent foundation for understanding Maple's language. The time you spend studying this chapter will save you hours puzzling over trivial problems in subroutines and procedures which appear to behave erratically. With the knowledge contained here, you should now see the source of such problems with clarity. Just as you may have done after finishing chapter 1, you may wish to put this book down for a while and practise creating more of your own procedures.

Chapter 3 introduces you to more advanced techniques in Maple programming. For example, it discusses procedures which return procedures, procedures which query the user for input, and packages which you can design yourself.

Read each of the remaining chapters in this book in any order. Feel free to skip to a chapter which discusses a topic in which you have an interest— perhaps the Maple debugger or programming with Maple graphics. If you wish a more formal presentation of the Maple language, take a look at chapters 4 and 5.

CHAPTER

3 Advanced Programming

As you progress in learning the Maple programming language and tackling more challenging projects, you may discover that you would like more detailed information. The topics in this chapter are more advanced than those in previous chapters, and some are difficult to follow without a sound understanding of Maple's evaluation rules, scoping rules, and other principal concepts.

The first two sections in this chapter begin where *Nested Procedures* on page 49 left off, using and returning procedures within the same procedure. Armed with a basic knowledge of Maple's evaluation rules, you will discover that such procedures are not difficult to write.

Surprisingly, local variables can exist long after the procedure which created them has finished. This feature can be particularly useful when you wish a procedure to return a procedure, but the new procedure needs a unique place to store information. Maple's assume facility, for example, uses such variables. The second section clearly explains and demonstrates how to use them effectively.

Three special topics make up the remainder of this chapter: interactive input, extending Maple, and creating your own packages. Interactive input allows you to write interactive procedures, making them more intuitive by querying the user for missing information. Perhaps you wish to write an interactive tutorial or a test. You are already aware of the customization power which you gain through the ability to write procedures; Maple also supplies some particularly useful mechanisms for modifying or extending Maple's functionality other than by writing a completely separate group of commands. The final section describes how to construct a package of procedures which operates like Maple's own packages, such as plots or linalg. In conjunction with the topics you find in the specialized chapters in the remainder of this book, the topics here will equip you to use Maple to its fullest.

3.1 Procedures Which Return Procedures

Of all the types of procedures you may want to write, procedures which return procedures are likely to cause the most trouble. Creating these procedures test whether you understand Maple's evaluation and scoping rules presented in chapter 2 where you learned about procedures within procedures, how Maple evaluates parameters, and assigns and evaluates both local and global variables. You also learned, for example, that an inner procedure will recognize the variable declarations of an outer procedure.

Some of the standard Maple commands return procedures. For example, rand returns a procedure which in turn produces randomly chosen integers from a specified range. If you use the type=numeric option with dsolve, it returns a procedure which supplies a numeric estimate of the solution to a differential equation.

You may wish to incorporate such features into your own programs. The areas which require your particular attention are those of conveying values from the outer procedure to the inner procedure, and the use of local variables to store information unique to a returned procedure. This section discusses the former. The latter is the topic of the next section, *When Local Variables Leave Home* on page 79.

Creating a Newton Iteration

Newton's method is one way of locating the roots of a function. First, you pick a point on the x-axis that you think might be close to a root. Next, you find the slope of the curve at the point you picked. Draw the tangent to the curve at that point and observe where the tangent intersects the x-axis. For most functions, this second point is closer to the real root than your initial guess. Thus, to find the root, all you need to do is use the new point as a new guess and keep drawing tangents and finding new points.

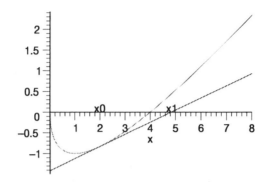

To find a numerical solution to the equation $f(x) = 0$, you may use Newton's method: guess an approximate solution, x_0; then use the following formula, which is the mathematical description of the above process, to generate better approximations.

$$x_{k+1} = x_k - \frac{f(x_k)}{f'(x_k)}$$

You can implement this algorithm on a computer in a number of ways. The program below takes a function and creates a new procedure, which takes an initial guess and, for that particular function, generates the next guess. Of course, the new procedure will not work for other functions. To find the roots of a new function, use MakeIteration to generate a new guess-generating procedure. The unapply command turns an expression into a procedure.

```
> MakeIteration := proc( expr::algebraic, x::name )
>    local iteration;
>    iteration := x - expr/diff(expr, x);
>    unapply(iteration, x);
> end:
```
Test the procedure on the expression $x - 2\sqrt{x}$.

```
> expr := x - 2*sqrt(x);
```

$$expr := x - 2\sqrt{x}$$

```
> Newton := MakeIteration( expr, x);
```

$$Newton := x \rightarrow x - \frac{x - 2\sqrt{x}}{1 - \frac{1}{\sqrt{x}}}$$

It only takes Newton a few iterations to find the solution, $x = 4$.

```
> x0 := 2.0;
```

$$x0 := 2.0$$

```
> to 4 do x0 := Newton(x0); od;
```

$$x0 := 4.828427124$$
$$x0 := 4.032533198$$
$$x0 := 4.000065353$$
$$x0 := 4.000000000$$

The MakeIteration procedure above expects its first argument to be an algebraic expression. You can also write a version of MakeIteration that works on functions. Since MakeIteration below knows that the parameter f is a procedure, you must use the eval command to evaluate it fully.

```
> MakeIteration := proc( f::procedure )
>    (x->x) - eval(f) / D(eval(f));
> end:
> g := x -> x - cos(x);
```

$$g := x \rightarrow x - \cos(x)$$

```
> SirIsaac := MakeIteration( g );
```

$$SirIsaac := (x \rightarrow x) - \frac{x \rightarrow x - \cos(x)}{x \rightarrow 1 + \sin(x)}$$

Note that SirIsaac does not contain references to the name g; thus, you can change g without breaking SirIsaac. You can find a good approximate solution to $x - \cos(x) = 0$ in a few iterations.

```
> x0 := 1.0;
```

$$x0 := 1.0$$

```
> to 4 do x0 := SirIsaac(x0) od;
```

$$x0 := .7503638679$$
$$x0 := .7391128909$$
$$x0 := .7390851334$$
$$x0 := .7390851332$$

A Shift Operator

Consider the problem of writing a procedure that takes a function, f, as input and returns a function, g, such that $g(x) = f(x + 1)$. You can write such a procedure in the following manner.

```
> shift := (f::procedure) -> ( x->f(x+1) ):
```

Try performing a shift on $\sin(x)$.

```
> shift(sin);
```

$$x \rightarrow \sin(x+1)$$

Maple's lexical scoping rules declare the f within the inner procedure to be the same f as the parameter within the outer procedure. Therefore, the command shift works as written.

The version of shift above works with univariate functions but it does not work with functions of two or more variables.

```
> h := (x,y) -> x*y;
```

$$h := (x, \ y) \rightarrow x\,y$$

```
> hh := shift(h);
```

$$hh := x \rightarrow h(x+1)$$

```
> hh(x,y);
```

```
Error, (in h) h uses a 2nd argument, y,
which is missing
```

If you want shift to work with multivariate functions, you must rewrite it to deal with the additional parameters. In a procedure, args is the sequence of actual parameters, and args[2..-1] is the sequence of actual parameters except the first one; see *Selection Operation* on page 146. It follows that the procedure x->f(x+1,args[2..-1]) passes all its arguments except the first directly to F.

```
> shift := (f::procedure) -> ( x->f(x+1, args[2..-1]) ):
```

```
> hh := shift(h);
```

$$hh := x \rightarrow h(x+1, \ args_{2..-1})$$

```
> hh(x,y);
```

$$(x+1)\,y$$

The function hh depends on h; if you change h, you implicitly change hh;

```
> h := (x,y,z) -> y*z^2/x;
```

$$h := (x, \ y, \ z) \rightarrow \frac{y\,z^2}{x}$$

```
> hh(x,y,z);
```

$$\frac{y\,z^2}{x+1}$$

3.2 When Local Variables Leave Home

Local or Global? on page 50 states that local variables are not only local to a procedure but also to an invocation of that procedure. Very simply, calling a procedure creates and uses new local variables each time. If you invoke the same procedure twice, the local variables it uses the second time are distinct from those it used the first time.

What may surprise you is that the local variables do not necessarily disappear when the procedure exits. You can write procedures which return a local variable, either explicitly or implicitly, to the interactive session, where it may survive indefinitely. You may find these renegade local variables confusing, particularly since they may have the same name as some global variables, or even other local variables which another procedure or a different call to the same procedure created. In fact, you can create as many distinct variables as you want, all with the same name.

The procedure below creates a new local variable, a, and then returns this new variable.

```
> make_a := proc()
>         local a;
>         a;
> end;
```

$$make\_a := \mathbf{proc}() \ \mathbf{local} \ a; \ a \ \mathbf{end}$$

Since a set in Maple contains *unique* elements, you can easily verify that each a that make_a returns is unique.

```
> test := { a, a, a };
```

$$test := \{a\}$$

```
> test := test union { make_a() };
```

$$test := \{a, \ a\}$$

```
> test := test union { 'make_a'()$5 };
```

$$test := \{a, \ a, \ a, \ a, \ a, \ a, \ a\}$$

Obviously, Maple identifies variables by more than their names.

Remember that no matter how many variables you create with the same name, when you type a name in an interactive session, Maple interprets that name to be of a *global* variable. Indeed, you can easily find the global a in the above set `test`.

```
> seq( evalb(i=a), i=test);
```

$$\text{true, false, false, false, false, false, false}$$

You can use local variables to make Maple print things it would not ordinarily be able to display. The above set `test` is an example. Another example is expressions which Maple would ordinarily simplify automatically. For example, Maple automatically simplifies the expression $a + a$ to $2a$, so displaying the equation $a + a = 2a$ is not easy. You can create the illusion that Maple is showing you these steps using procedure `make_a`, above.

```
> a + make_a() = 2*a;
```

$$a + a = 2\,a$$

To Maple, these two variables are distinct, even though they share the same name.

You cannot easily assign a value to such escapees. Whenever you type a name in an interactive session, Maple thinks you mean the global variable of that name. While this prevents you from using the assignment *statement*, it does not prevent you from using the assignment *command*. The trick is to write a Maple expression which extracts the variable you want. For example, in the equation above, you may extract the local a by removing the global a from the left-hand side of the equation.

```
> eqn := %;
```

$$eqn := a + a = 2\,a$$

```
> another_a := remove( x->evalb(x=a), lhs(eqn) );
```

$$another\_a := a$$

You may then assign the global name a to this extracted variable and so verify the equation.

```
> assign(another_a = a);
> eqn;
```

$$2\,a = 2\,a$$

```
> evalb(%);
```

$$\text{true}$$

Should your expression be complicated, you may need a fancier command to extract the desired variable.

You may have encountered this situation before without realizing it, when you were using the assume facility and wished to remove an assumption. The assume facility attaches various definitions to the variable you specify, with one result being that the name subsequently appears as a *local* name with an appended tilde. Maple does not understand if you type the tilded name because no relationship exists with the *global* variable name containing a tilde.

```
> assume(b>0);
> x := b + 1;
```

$$x := b\tilde{} + 1$$

```
> subs( 'b~'=c, x);
```

$$b\tilde{} + 1$$

When you clear the definition of the named variable the association between the name and the local name with the tilde is lost, but expressions created with the local name still contain it.

```
> b := evaln(b);
```

$$b := b$$

```
> x;
```

$$b\tilde{} + 1$$

If you later wish to reuse your expression, you must either perform a substitution before removing the assumption or perform some manipulations of your expressions similar to the equation eqn above.

Creating the Cartesian Product of a Sequence of Sets

An important use for returning local objects arises when the returned object is a procedure. When you write a procedure which returns a procedure, you will often find it useful to have the procedure create a variable which holds information pertinent only to the returned procedure. This allows different procedures (or different invocations of the same procedure) to pass information between themselves.

The program introduced in this section uses this idea. When you pass a sequence of sets to the procedure it constructs a new procedure. The new procedure returns the next term in the Cartesian product each time you invoke it. Local variables from the outer procedure are used to keep track of which term to return next.

The *Cartesian product* of a sequence of sets is the set of all lists whose ith entry is an element of the ith set. Thus, the Cartesian product of $\{\alpha, \beta, \gamma\}$

and $\{x, y\}$ is

$$\{\alpha, \beta, \gamma\} \times \{x, y\} = \{[\alpha, x], [\beta, x], [\gamma, x], [\alpha, y], [\beta, y], [\gamma, y]\}.$$

The number of elements in the Cartesian product of a sequence of sets grows very rapidly as the sequence gets longer or the sets get larger. It therefore requires a large amount of memory to store all the elements of the Cartesian product. One way around this is to write a procedure that returns a new element of the Cartesian product each time you call it. By calling such a procedure repeatedly you can process every element in the Cartesian product without ever storing all its elements at once.

The procedure below returns the next element of the Cartesian product of the list s of sets. It uses an array, c, of counters to keep track of which element comes next. For example, c[1]=3 and c[2]=1 correspond to the third element of the first set and the first element of the second set.

```
> s := [ {alpha, beta, gamma}, {x, y} ];
```

$$s := [\{\gamma, \beta, \alpha\}, \{x, y\}]$$

```
> c := array( 1..2, [3, 1] );
```

$$c := [3, 1]$$

```
> [ seq( s[j][c[j]], j=1..2 ) ];
```

$$[\alpha, x]$$

Before you call the element procedure you must initialize all the counters to 1, except the first one, which should be 0.

```
> c := array( [0, 1] );
```

$$c := [0, 1]$$

In element below, nops(s) is the number of sets and nops(s[i]) is the number of elements in the ith set. When you have seen all the elements, the procedure re-initializes the array of counters and returns FAIL. Therefore, you can trace through the Cartesian product again by calling element again.

```
> element := proc(s::list(set), c::array(1, nonnegint))
>     local i, j;
>     for i to nops(s) do
>         c[i] := c[i] + 1;
>         if c[i] <= nops( s[i] ) then
>             RETURN( [ seq(s[j][c[j]], j=1..nops(s)) ] );
>         fi;
>         c[i] := 1;
```

```
>     od;
>     c[1] := 0;
>     FAIL;
> end:
> element(s, c); element(s, c); element(s, c);
```

$$[\gamma, x]$$
$$[\beta, x]$$
$$[\alpha, x]$$

```
> element(s, c); element(s, c); element(s, c);
```

$$[\gamma, y]$$
$$[\beta, y]$$
$$[\alpha, y]$$

```
> element(s, c);
```

FAIL

Instead of writing a new procedure for each Cartesian product you want to study, you can write a procedure, CartesianProduct, that returns such a procedure. CartesianProduct below first creates a list, s, of its arguments, which should all be sets. Then it initializes the array, c, of counters and defines the subprocedure element. Finally, the element subprocedure is invoked inside a proc structure.

```
> CartesianProduct := proc()
>     local s, c, element;
>     global S, C;
>     s := [args];
>     if not type(s, list(set)) then
>         ERROR( "expected a sequence of sets, but received",
>                 args );
>     fi;
>     c := array( [0, 1$(nops(s)-1)] );
>
>     element := proc(s::list(set), c::array(1, nonnegint))
>         local i, j;
>         for i to nops(s) do
>             c[i] := c[i] + 1;
>             if c[i] <= nops( s[i] ) then
>                 RETURN( [ seq(s[j][c[j]], j=1..nops(s)) ] );
>             fi;
>             c[i] := 1;
>         od;
>         c[1] := 0;
```

```
>         FAIL;
>     end;
>
>     proc()
>         element(s, c);
>     end;
> end:
```

Again, you can find all six elements of $\{\alpha, \beta, \gamma\} \times \{x, y\}$.

```
> f := CartesianProduct( {alpha, beta, gamma}, {x,y} );
```

$$f := \textbf{proc}()\ element(s,\ c)\ \textbf{end}$$

```
> to 7 do f() od;
```

$$[\gamma,\ x]$$
$$[\beta,\ x]$$
$$[\alpha,\ x]$$
$$[\gamma,\ y]$$
$$[\beta,\ y]$$
$$[\alpha,\ y]$$
$$FAIL$$

You can use `CartesianProduct` to study several products simultaneously.

```
> g := CartesianProduct( {x, y}, {N, Z, R},
>                         {56, 23, 68, 92} );
```

$$g := \textbf{proc}()\ element(s,\ c)\ \textbf{end}$$

The following are the first few elements of $\{x, y\} \times \{N, Z, R\} \times \{56, 23, 68, 92\}$.

```
> to 5 do g() od;
```

$$[x,\ R,\ 56]$$
$$[y,\ R,\ 56]$$
$$[x,\ Z,\ 56]$$
$$[y,\ Z,\ 56]$$
$$[x,\ N,\ 56]$$

Because the variables s in f and g are local variables to `CartesianProduct`, they are not shared by different *invocations* of `CartesianProduct`. Similarly, the variable c in f and g is also not shared. You can see that the two arrays of counters are different by invoking f and g a few more times.

```
> to 5 do f(), g() od;
```

$$[\gamma, x], [y, N, 56]$$
$$[\beta, x], [x, R, 68]$$
$$[\alpha, x], [y, R, 68]$$
$$[\gamma, y], [x, Z, 68]$$
$$[\beta, y], [y, Z, 68]$$

The element procedure in g is also local to CartesianProduct. Therefore, you can change the value of the global variable element without breaking g.

```
> element := 45;
```

$$element := 45$$

```
> g();
```

$$[x, N, 68]$$

These examples demonstrate not only that local variables can escape the bounds of the procedures which create them, but that this mechanism allows you to write procedures which create specialized procedures.

Exercises

1. The procedure that CartesianProduct generates does not work if one of the sets is empty.

```
> f := CartesianProduct( {}, {x,y} );
```

$$f := \mathbf{proc}() \; element(s, \; c) \; \mathbf{end}$$

```
> f();
```

```
Error, (in element) invalid subscript selector
```

Improve the type-checking in CartesianProduct so that it generates an informative error message in each such case.

2. A *partition* of a positive integer, *n*, is a list of positive integers whose sum is *n*. The same integer can appear several times in the partition but the order or the integers in the partition is irrelevant. Thus, the following are all the partitions of 5:

$$[1, 1, 1, 1, 1], [1, 1, 1, 2], [1, 1, 3], [1, 2, 2], [1, 4], [2, 3], [5].$$

Write a procedure that generates a procedure that returns a new partition of *n* each time you call it.

3.3 Interactive Input

Normally you pass input to Maple procedures as parameters. Sometimes, however, you may want to write a procedure that asks the user directly for input. For example, you could write a procedure that drills students on some topic; the procedure could generate random problems and verify the students' answers. The input may be the value of a certain parameter, or the answer to a question such as whether a parameter is positive or not. The two commands in Maple for reading input from the terminal are the `readline` command and the `readstat` command.

Reading Strings from the Terminal

The `readline` command reads one line of text from a file or the keyboard. You may use the `readline` command as follows.

```
readline( filename )
```

If *filename* is the special name `terminal`, then `readline` reads a line of text from the keyboard. `readline` returns the text as a string.

```
> s := readline( terminal );
```

Waterloo Maple, Inc.

$$s := \text{``Waterloo Maple, Inc.''}$$

Here is a simple application, prompting the user for an answer to a question

```
> DetermineSign := proc(a::algebraic) local s;
>     printf("Is the sign of %a positive?  Answer yes or no: ",a);
>     s := readline(terminal);
>     evalb( s="yes" or s = "y" );
> end:

> DetermineSign(u-1);
```

Is the sign of u-1 positive? Answer yes or no: y

$$true$$

Reading Text Lines from a File on page 340 gives more details on the `readline` command.

Reading Expressions from the Terminal

You may want to write procedures that require the user to input an expression rather than a string. The readstat command reads one expression from the keyboard.

```
readstat( prompt )
```

The *prompt* is an optional string.

```
> readstat("Enter degree: ");

Enter degree: n-1;
```

$$n - 1$$

Notice that the readstat command insists on a terminating semicolon (or colon). Unlike the readline command, which only reads one line, the readstat command works like the rest of Maple: it allows you to break a large expression across multiple lines. Another advantage of using the readstat command is that if the user makes a mistake in the input, the readstat command will automatically re-prompt the user for input, giving the user an opportunity to correct the error.

```
> readstat("Enter a number: ");

Enter a number: 5^^8;
syntax error, '^' unexpected:
5^^8;
   ^

Enter a number: 5^8;
```

$$390625$$

Here is an application of the readstat command for implementing an interface to the limit command. The idea is: given the function $f(x)$, assume x is the variable if only one variable is present; otherwise, ask the user what the variable is, and also ask the user for the limit point.

```
> GetLimitInput := proc(f::algebraic)
>     local x, a, I;
>     # choose all variables in f
>     I := select(type, indets(f), name);
>
>     if nops(I) = 1 then
>         x := I[1];
>     else
>         x := readstat("Input limit variable: ");
>         while not type(x, name) do
```

```
>               printf("A variable is required: received %a\n", x);
>               x := readstat("Please re-input limit variable: ");
>          od;
>     fi;
>     a := readstat("Input limit point: ");
>     x = a;
> end:
```

The expression $\sin(x)/x$ depends only on one variable, so `GetLimitInput` does not ask for any limit variable.

```
> GetLimitInput( sin(x)/x );

Input limit point: 0;
```

$$x = 0$$

Below, the user first tries to use the number 1 as the limit variable. Since 1 is not a name, `GetLimitInput` asks for another limit variable.

```
> GetLimitInput( exp(u*x) );

Input limit variable: 1;
A variable is required: received 1
Please re-input limit variable: x;
Input limit point: infinity;
```

$$x = \infty$$

You can specify a number of options to `readstat`; see *Reading Maple Statements* on page 345.

Converting Strings to Expressions

Some times you may need more control over how and when Maple evaluates user input to your procedure than the `readstat` command allows. In such cases, you can use the `readline` command to read the input as a string, and the `parse` command to convert the string to an expression. The string must represent a complete expression.

```
> s := "a*x^2 + 1";
```

$$s := \text{``}a*x\hat{}2 + 1\text{''}$$

```
> y := parse( s );
```

$$y := a\,x^2 + 1$$

When you parse the string s you get an expression. In this case, you get a sum.

```
> type(s, string), type(y, '+');
```

$$true, \; true$$

The parse command does not evaluate the expression it returns. You must use eval to evaluate the expression explicitly. Below, Maple does not evaluate the variable a to its value, 2, until you explicitly use the eval command.

```
> a := 2;
```

$$a := 2$$

```
> z := parse( s );
```

$$z := a\,x^2 + 1$$

```
> eval(z);
```

$$2\,x^2 + 1$$

See *Parsing Maple Expressions and Statements* on page 360 for more details about the parse command.

The techniques you have seen in this section are all very simple, but you can use them to create powerful applications such as Maple tutorials, procedures that drill students, or interactive lessons.

3.4 Extending Maple

Even though you may find it useful to write your own procedures to perform new tasks, sometimes extending the abilities of Maple's own commands is most beneficial. Many of Maple's existing commands provide this service. This section familiarizes you with the most helpful methods, including making your own types and operators, modifying how Maple displays expressions, and extending the abilities of such useful commands as simplify and expand.

Defining New Types

If you use a complicated structured type you may find it easier to assign the structured type to a variable of the form 'type/*name*'. That way you only have to write the structure once, thus reducing the risk of errors. When you have defined the variable 'type/*name*', you can use *name* as a type.

```
> 'type/Variables' := {name, list(name), set(name)}:
```

```
> type( x, Variables );
```

true

```
> type( { x[1], x[2] }, Variables );
```

true

When the structured type mechanism is not powerful enough, you can define a new type by assigning a procedure to a variable of the form 'type/*name*'. When you test whether an expression is of type *name*, Maple invokes the procedure 'type/*name*' on the expression if such a procedure exists. Your procedure should return true or false. The 'type/permutation' procedure below determines if *p* is a permutation of the first *n* positive integers. That is, *p* should contain exactly one copy of each integer from 1 through *n*.

```
> 'type/permutation' := proc(p)
>    local i;
>    type(p,list) and { op(p) } = { seq(i, i=1..nops(p)) };
> end:
> type( [1,5,2,3], permutation );
```

false

```
> type( [1,4,2,3], permutation );
```

true

Your type-testing procedure may have more than one parameter. When you test if an expression, *expr*, has type *name*(*parameters*), then Maple invokes

```
'type/name'( expr, parameters )
```

if such a procedure exists. The 'type/LINEAR' procedure below determines if *f* is a polynomial in *V* of degree 1.

```
> 'type/LINEAR' := proc(f, V::name)
>    type( f, polynom(anything, V) ) and degree(f, V) = 1;
> end:

> type( a*x+b, LINEAR(x) );
```

true

```
> type( x^2, LINEAR(x) );
```

false

```
> type( a, LINEAR(x) );
```

$$false$$

Exercises

1. Modify the 'type/LINEAR' procedure so that you can use it to test if an expression is linear in a set of variables. For example, $x + ay + 1$ is linear in both x and y, but $xy + a + 1$ is not.
2. Define the type POLYNOM(X) which tests if an algebraic expression is a polynomial in X where X may be a name, a list of names, or a set of names.

Neutral Operators

Maple knows a number of operators, for example +, *, ^, and, not, and union. All these operators have special meaning to Maple: they represent algebraic operations, such as addition or multiplication, or logical operations, or operations on sets. Maple also has a special class of operators, the *neutral operators*, on which it does not impose any meaning. Instead Maple allows *you* to define the meaning of any neutral operator. The name of a neutral operator begins with the ampersand character, &. *The Neutral Operators* on page 157 describes the naming conventions for neutral operators.

```
> 7 &^ 8 &^ 9;
```

$$(7 \,\&^\wedge 8) \,\&^\wedge 9$$

```
> evalb( 7 &^ 8 = 8 &^ 7 );
```

$$false$$

```
> evalb( (7&^8)&^9 = 7&^(8&^9) );
```

$$false$$

Internally, Maple represents neutral operators as procedure calls; thus, 7&^8 is just a convenient way of writing &^(7,8).

```
> &^(7, 8);
```

$$7 \,\&^\wedge 8$$

Maple only uses the infix notation if your neutral operator has exactly two arguments.

```
> &^(4), &^(5, 6), &^(7, 8, 9);
```

$$\&^\wedge(4), \; 5 \,\&^\wedge 6, \; \&^\wedge(7, 8, 9)$$

You can define the actions of a neutral operator by assigning a procedure to its name. The example below implements the Hamiltonians by assigning a neutral operator to a procedure that multiplies two Hamiltonians. The next paragraph explains all you need to know about the Hamiltonians to understand the example.

The *Hamiltonians* or *Quaternians* extend the complex numbers in the same way the complex numbers extend the real numbers. Each Hamiltonian has the form $a + bi + cj + dk$ where a, b, c, and d are real numbers. The special symbols i, j, and k satisfy the following multiplication rules: $i^2 = -1$, $j^2 = -1$, $k^2 = -1$, $ij = k$, $ji = -k$, $ik = -j$, $ki = j$, $jk = i$, and $kj = -i$.

The '&^' procedure below uses I, J, and K as the three special symbols. Therefore, you should remove the alias from I so that it no longer denotes the *complex* imaginary unit.

```
> alias( I=I );
```

You can multiply many types of expressions using '&^', making it convenient to define a new type, Hamiltonian, by assigning a structured type to the name 'type/Hamiltonian'.

```
> 'type/Hamiltonian' := { '+', '*', name, realcons,
>     specfunc(anything, '&^') };
```

$$\text{'type/Hamiltonian'} :=$$

$$\{\text{'*'}, \text{'+'}, \textit{name}, \textit{realcons}, \text{specfunc}(\textit{anything}, \text{'\& ^'})\}$$

The '&^' procedure multiplies the two Hamiltonians, x and y. If either x or y is a real number or variable, then their product is the usual product denoted by * in Maple. If x or y is a sum, '&^' maps the product onto the sum; that is, '&^' applies the distributive laws: $x(u + v) = xu + xv$ and $(u+v)x = ux+vx$. If x or y is a product, '&^' extracts any real factors. You must take special care to avoid infinite recursion when x or y is a product that does not contain any real factors. If none of the multiplication rules apply, '&^' returns the product unevaluated.

```
> '&^' := proc( x::Hamiltonian, y::Hamiltonian )
>    local Real, unReal, isReal;
>    isReal := z -> evalb( is(z, real) = true );
>
>    if isReal(x) or isReal(y) then
>        x * y;
>
>    elif type(x, '+') then
>        # x is a sum, u+v, so x&^y = u&^y + v&^y.
```

```
>         map('&^', x, y);
>
>     elif type(y, '+') then
>         # y is a sum, u+v, so x&^y = x&^u + x&^v.
>         map2('&^', x, y);
>
>     elif type(x, '*') then
>         # Pick out the real factors of x.
>         Real := select(isReal, x);
>         unReal := remove(isReal, x);
>         # Now x&^y = Real * (unReal&^y)
>         if Real=1 then
>             if type(y, '*') then
>                 Real := select(isReal, y);
>                 unReal := remove(isReal, y);
>                 Real * '&^'(x, unReal);
>             else
>                 '&^'(x, y);
>             fi;
>         else
>             Real * '&^'(unReal, y);
>         fi;
>
>     elif type(y, '*') then
>         # Similar to the x-case but easier since
>         # x cannot be a product here.
>         Real := select(isReal, y);
>         unReal := remove(isReal, y);
>         if Real=1 then
>             '&^'(x, y);
>         else
>             Real * '&^'(x, unReal);
>         fi;
>
>     else
>         '&^'(x,y);
>     fi;
> end:
```

You can place all the special multiplication rules for the symbols *I*, *J*, and *K* in the remember table of `&^`. See *Remember Tables* on page 68.

```
> '&^'(I,I) := -1: '&^'(J,J) := -1: '&^'(K,K) := -1:
> '&^'(I,J) := K: '&^'(J,I) := -K:
> '&^'(I,K) := -J: '&^'(K,I) := J:
```

```
> '&^'(J,K) := I: '&^'(K,J) := -I:
```

Since '&^' is a neutral operator, you can write products of Hamiltonians using &^ as the multiplication symbol.

```
> (1 + 2*I + 3*J + 4*K) &^ (5 + 3*I - 7*J);
```

$$20 + 41\,I + 20\,J - 3\,K$$

```
> (5 + 3*I - 7*J) &^ (1 + 2*I + 3*J + 4*K);
```

$$20 - 15\,I - 4\,J + 43\,K$$

```
> 56 &^ I;
```

$$56\,I$$

Below, a is an unknown Hamiltonian until you tell Maple that a is an unknown real number.

```
> a &^ J;
```

$$a\,\&^{\hat{}}\,J$$

```
> assume(a, real);
> a &^ J;
```

$$a\tilde{\ }\,J$$

Exercise

1. The inverse of a general Hamiltonian, $a + bi + cj + dk$, is $(a - bi - cj - dk)/(a^2 + b^2 + c^2 + d^2)$. You can demonstrate that fact as follows: Assume that $a, b, c,$ and d are real and define a general Hamiltonian, h.

```
> assume(a, real); assume(b, real);
> assume(c, real); assume(d, real);
> h := a + b*I + c*J + d*K;
```

$$h := a\tilde{\ } + b\tilde{\ }I + c\tilde{\ }J + d\tilde{\ }K$$

By the formula above, the following should be the inverse of h.

```
> hinv := (a-b*I-c*J-d*K) / (a^2+b^2+c^2+d^2);
```

$$hinv := \frac{a\tilde{\ } - b\tilde{\ }I - c\tilde{\ }J - d\tilde{\ }K}{a\tilde{\ }^2 + b\tilde{\ }^2 + c\tilde{\ }^2 + d\tilde{\ }^2}$$

Now all you have to check is that h &^ hinv and hinv &^ h both simplify to 1.

> h &^ hinv;

$$\frac{a\tilde{}\,(a\tilde{}\,-b\tilde{}\,I-c\tilde{}\,J-d\tilde{}\,K)}{\%1}$$

$$+\frac{b\tilde{}\,(I\,a\tilde{}\,+b\tilde{}\,-c\tilde{}\,K+d\tilde{}\,J)}{\%1}$$

$$+\frac{c\tilde{}\,(J\,a\tilde{}\,+b\tilde{}\,K+c\tilde{}\,-d\tilde{}\,I)}{\%1}$$

$$+\frac{d\tilde{}\,(K\,a\tilde{}\,-b\tilde{}\,J+c\tilde{}\,I+d\tilde{})}{\%1}$$

$$\%1 := a\tilde{}^2 + b\tilde{}^2 + c\tilde{}^2 + d\tilde{}^2$$

> simplify(%);

$$1$$

> hinv &^ h;

$$\frac{a\tilde{}\,(a\tilde{}\,-b\tilde{}\,I-c\tilde{}\,J-d\tilde{}\,K)}{\%1}$$

$$+\frac{a\tilde{}\,b\tilde{}\,I+b\tilde{}^2 + b\tilde{}\,c\tilde{}\,K - b\tilde{}\,d\tilde{}\,J}{\%1}$$

$$+\frac{a\tilde{}\,c\tilde{}\,J - b\tilde{}\,c\tilde{}\,K + c\tilde{}^2 + c\tilde{}\,d\tilde{}\,I}{\%1}$$

$$+\frac{a\tilde{}\,d\tilde{}\,K + b\tilde{}\,d\tilde{}\,J - c\tilde{}\,d\tilde{}\,I + d\tilde{}^2}{\%1}$$

$$\%1 := a\tilde{}^2 + b\tilde{}^2 + c\tilde{}^2 + d\tilde{}^2$$

> simplify(%);

$$1$$

Write a procedure, '&/', that computes the inverse of a Hamiltonian. You may want to implement the following rules.

$$\&/(\ \&/x\ ) = x, \quad \&/(x\&^\wedge y) = (\&/y) \ \&^\wedge \ (\&/x),$$
$$x \ \&^\wedge \ (\&/x) = 1 = (\&/x) \ \&^\wedge \ x.$$

Extending Certain Commands

If you introduce your own data structures, Maple cannot know how to manipulate them. In most cases, you design new data structures because you want to write special-purpose procedures that manipulate them, but sometimes extending the capabilities of one or more of Maple's built-in commands is more intuitive. You can extend several Maple commands, among them expand, simplify, diff, series, and evalf.

Suppose you choose to represent a polynomial $a_n u^n + a_{n-1}u^{n-1} + \cdots + a_1 u + a_0$ using the data structure

```
POLYNOM( u, a_0, a_1, ..., a_n )
```

You can then extend the diff command so that you can differentiate polynomials represented in that way. If you write a procedure with a name of the form `diff/F` then diff invokes it on any unevaluated calls to F. Specifically, if you use diff to differentiate F(*arguments*) with respect to x, then diff invokes `diff/F` as follows.

```
'diff/F'( arguments, x )
```

The procedure below differentiates a polynomial in u with constant coefficients with respect to x.

```
> 'diff/POLYNOM' := proc(u)
>     local i, s, x;
>     x := args[-1];
>     s := seq( i*args[i+2], i=1..nargs-3 );
>     'POLYNOM'(u, s) * diff(u, x);
> end:
> diff( POLYNOM(x, 1, 1, 1, 1, 1, 1, 1, 1, 1, 1), x );
```

$$POLYNOM(x, 1, 2, 3, 4, 5, 6, 7, 8, 9)$$

```
> diff( POLYNOM(x*y, 34, 12, 876, 11, 76), x );
```

$$POLYNOM(x\,y, 12, 1752, 33, 304)\,y$$

The implementation of the Hamiltonians that *Neutral Operators* on page 91 describes does not know that multiplication of Hamiltonians is associative, that is $(xy)z = x(yz)$. Sometimes, using associativity simplifies a result. Recall that I here is *not* the complex imaginary unit, but rather, one of the special symbols I, J, and K that are part of the definition of the Hamiltonians.

```
> x &^ I &^ J;
```

$$(x \,\&^{\hat{}}\, I) \,\&^{\hat{}}\, J$$

```
> x &^ ( I &^ J );
```

$$x \mathbin{\&^{\hat{}}} K$$

You can extend the `simplify` command so that it applies the associative law to unevaluated products of Hamiltonians. If you write a procedure with a name of the form `'simplify/F'`, then `simplify` invokes it on any unevaluated function calls to *F*. Thus, you must write a procedure `'simplify/&^'` that applies the associative law to Hamiltonians.

The procedure below uses the `typematch` command to determine if its argument is of the form (a&^b)&^c and, if so, it picks out the a, b, and c.

```
> s := x &^ y &^ z;
```

$$s := (x \mathbin{\&^{\hat{}}} y) \mathbin{\&^{\hat{}}} z$$

```
> typematch( s, ''&^''( ''&^''( a::anything, b::anything ),
>                       c::anything ) );
```

true

```
> a, b, c;
```

$$x, \, y, \, z$$

You can give the user details of which simplifications your procedure makes through the `userinfo` command. The `'simplify/&^'` procedure prints out an informative message if you set `infolevel[simplify]` or `infolevel[all]` to at least 2.

```
> 'simplify/&^' := proc( x )
>    local a, b, c;
>    if typematch( x,
>          ''&^''( ''&^''( a::anything, b::anything ),
>                  c::anything ) ) then
>       userinfo(2, simplify, "applying the associative law");
>       a &^ ( b &^ c );
>    else
>       x;
>    fi;
> end:
```

Applying the associative law does make some products of Hamiltonians simpler.

```
> x &^ I &^ J &^ K;
```

$$((x \mathbin{\&^{\hat{}}} I) \mathbin{\&^{\hat{}}} J) \mathbin{\&^{\hat{}}} K$$

```
> simplify(%);
```

$$-x$$

If you set `infolevel[simplify]` large enough, Maple prints out information on what `simplify` tries in order to make your expression simpler.

```
> infolevel[simplify] := 5;
```

$$infolevel_{simplify} := 5$$

```
> w &^ x &^ y &^ z;
```

$$((w \,\&^{\wedge} x) \,\&^{\wedge} y) \,\&^{\wedge} z$$

```
> simplify(%);
```

```
simplify:    applying    &^   function to expression
simplify:    applying    &^   function to expression
simplify/&^:   "applying the associative law"
simplify:    applying    &^   function to expression
simplify/&^:   "applying the associative law"
```

$$w \,\&^{\wedge} ((x \,\&^{\wedge} y) \,\&^{\wedge} z)$$

The help pages for `expand`, `series`, and `evalf` provide details on how you may extend those commands. See also *Extending the* `evalf` *Command* on page 251.

You may employ any or all of the above methods, as you see fit. Maple's design affords you the opportunity to customize it to suit your needs, allowing you great flexibility.

3.5 Writing Your Own Packages

You will often find yourself writing a set of procedures which are all related. If you intend to use all of them on one particular type of problem, it makes sense to group these commands together. This type of situation arises frequently within Maple. For this reason, the Maple designers have grouped some types of commands, such as those for drawing plots or those for working on problems in linear algebra, together in the form of packages. You can also quite easily make packages of your own procedures.

Maple views a package as a type of table. Each key in the table is the name of a procedure, and the value corresponding to the key is the procedure. All you have to do is define the table, just as you would define any Maple table, then save it in a `.m` file. Maple can easily recognize your package and treat it like any of the ones with which it is already familiar.

One advantage of saving a group of related procedures as a table is that the user can load the package using the `with` command. A *package* is simply a table of procedures.

```
> powers := table();
```

$$powers := \text{table}([$$

$$])$$

```
> powers[sqr1] := proc(x::anything)
>     x^2;
> end:
> powers[cube] := proc(x::anything)
>     x^3;
> end:
```

You can now square any expression by calling powers[sqr1].

```
> powers[sqr1](x+y);
```

$$(x + y)^2$$

Below, the single quotes around sqr1 ensure that powers[fourth] works even if you assign a value to sqr1.

```
> powers[fourth] := proc(x::anything)
>     powers['sqr1']( powers['sqr1'](x) );
> end:
> sqr1 := 56^2;
```

$$sqr1 := 3136$$

```
> powers[fourth](x);
```

$$x^4$$

```
> sqr1 := 'sqr1':
```

You can use the with command to define all the short names for the procedures in your package.

```
> with(powers);
```

$$[cube, fourth, sqr1]$$

```
> cube(x);
```

$$x^3$$

Since your package is a table, you can save it as you would save any other Maple object. The command below saves powers in Maple's binary format to the file powers.m in the directory /users/yourself/tmp1.

```
> save( powers, "/users/yourself/tmp1/powers.m" );
```

```
> restart;
```

The with command can load your new package, provided you save it in a file with a name of the form *packagename* . m, and provided you tell Maple in which directories it should search for the file. The variable libname is a sequence of directory names which Maple searches in order from left to right. Thus you must prepend the name of the directory containing your package to libname.

```
> libname := "/users/yourself/tmp1", libname;
```

$$libname :=$$

"/users/yourself/tmp1", "/maple/lib"

You can now load the powers package and commence using it.

```
> with(powers);
```

[*cube, fourth, sqr1*]

```
> cube(3);
```

27

Although this package may seem rather silly, it demonstrates how easily you can make one. Define a table of procedures; save the table to a file whose name is that of the package with .m appended; and modify libname so that Maple will look in the directory where you saved the file.

You now know the basics and indeed may wish to suspend your reading and make a package or two. The rest of the chapter discusses the finer points of package creation, including initialization files and making your own Maple library.

Package Initialization

Some packages require access to specialized support code, such as special purpose types, formatted printing, or customized extensions on certain Maple commands. You can define such support code in an initialization procedure in your package. When the with command loads a package, it invokes the package procedure init, if it exists.

Maple already supports computations with complex numbers. However, for the purposes of example, suppose you want to write a package, cmplx, for complex number arithmetic that represents the complex number $a + bi$, where a and b are real, as COMPLEX(a, b). Thus, you must define a new data type, COMPLEX.

```
> 'type/COMPLEX' := 'COMPLEX'(realcons, realcons);
```

$$\text{'type/COMPLEX'} := \text{COMPLEX}(\textit{realcons, realcons})$$

```
> z := COMPLEX(3, 4);
```

$$z := \text{COMPLEX}(3,\ 4)$$

```
> type(z, COMPLEX);
```

$$\textit{true}$$

You can easily pick out the real and imaginary parts of z: they are the first and second operands of z, respectively.

```
> cmplx[realpart] := proc(z::COMPLEX)
>    op(1, z);
> end:
> cmplx[imagpart] := proc(z::COMPLEX)
>    op(2, z);
> end:
> cmplx[realpart](z);
```

$$3$$

If you have the real and imaginary parts of a complex number, you can put them together in a COMPLEX structure.

```
> cmplx[makecomplex] := proc(a::realcons, b::realcons)
>    'COMPLEX'(a, b);
> end:
> w := cmplx[makecomplex](6, 2);
```

$$w := \text{COMPLEX}(6,\ 2)$$

You add two complex numbers by adding the real and imaginary parts separately.

```
> cmplx[addition] := proc(z::COMPLEX, w::COMPLEX)
>    local x1, x2, y1, y2;
>    x1 := cmplx['realpart'](z);
>    y1 := cmplx['imagpart'](z);
>    x2 := cmplx['realpart'](w);
>    y2 := cmplx['imagpart'](w);
>    cmplx['makecomplex'](x1+x2, y1+y2);
> end:
> cmplx[addition](z, w);
```

$$\text{COMPLEX}(9,\ 6)$$

The commands in your package rely on the global variable, `'type/COMPLEX'`, so you must define that variable in the initialization procedure of the package.

```
> cmplx[init] := proc()
>    global 'type/COMPLEX';
>    'type/COMPLEX' := 'COMPLEX'(realcons, realcons);
> end:
```

In this configuration, the `cmplx` package must be loaded using the `with` command in order for `'type/COMPLEX'` (and the functions that rely on it) to work. You can leave the `cmplx[init]` procedure out of the package if you save `'type/COMPLEX'` separately from `cmplx`.

Now you are ready to save the package. You should call the file `cmplx.m` so that `with` can load it.

```
> save( cmplx, "/users/yourself/tmp1/cmplx.m" );
```

You must prepend the directory containing your packages to `libname` so that `with` can find them.

```
> restart;
> libname := "/users/yourself/tmp1", libname;
```

$$libname :=$$

$$\text{"/users/yourself/tmp1", "/maple/lib"}$$

```
> with(cmplx);
```

$$[addition, imagpart, init, makecomplex, realpart]$$

The initialization procedure defines the `COMPLEX` type.

```
> type( makecomplex(4,3), COMPLEX );
```

$$true$$

Making Your Own Library

Instead of saving Maple objects to files in the directory structure on your computer, you may combine the contents of these files into one bigger file by creating your own Maple *library*. Using a library allows you to use file names that are independent of your computer's operating system. Thus, libraries make it easier to share your procedures with friends who use different kinds of computers. Libraries also tend to consume less disk space than when the same information is saved in many smaller files. A directory can contain at most one Maple library. A Maple library can contain arbitrarily many files; indeed, Maple comes with only one library that contains all the library-defined functions, including those in all the packages.

Before you make your own library *you should write-protect the main Maple library to ensure that you do not corrupt it by accident*. The variable `libname` is a sequence of the directories that contain your Maple libraries. The following is the value of `libname` on the computer that typeset these pages; the value will be different on your computer.

```
> libname;
```

$$\text{``/maple/lib''}$$

You cannot create a new library from within Maple. You must use the program *March* that comes with Maple. Exactly how you invoke *March* varies with the operating system—check your platform-specific documentation. On UNIX, the following command creates a new library in the directory /users/yourself/tmp1 which initially can contain ten `.m` files. march -c /users/yourself/tmp1 10 The Maple help page ?march explains *March* in more detail. Once you have created a new library, you must tell Maple where to find it by prepending the directory to `libname`.

```
> libname := "/users/yourself/tmp1", libname;
```

$$libname :=$$

$$\text{``/users/yourself/tmp1''}, \quad \text{``/maple/lib''}$$

You can use the `savelib` command to save any Maple object to a file in a library as follows.

```
savelib( nameseq, "file" )
```

Here, *nameseq* is a sequence of the names of the objects you want to save in *file* in a library. You may save several objects in a file whether or not the file is in a library. For files in libraries, Maple uses the convention that the file *name*`.m` must define the object *name*. The variable `savelibname` tells `savelib` in which directory your library is located.

```
> savelibname := "/users/yourself/tmp1";
```

$$savelibname := \text{``/users/yourself/tmp1''}$$

Note that you must also list the directory in `savelibname` in `libname`. The commands below define the powers package and saves it to your new library.

```
> powers := table():
```

```
> powers[sqr1] := proc(x::anything)
>    x^2;
> end:
> powers[cube] := proc(x::anything)
>    x^3;
> end:
> powers[fourth] := proc(x::anything)
>    powers['sqr1']( powers['sqr1'](x) );
> end:
> savelib( 'powers', "powers.m" );
```

The with command can also load packages from your own library provided libname contains the directory that contains your library.

```
> restart;
> libname := "/users/yourself/tmp1", libname;
```

$$libname :=$$

$$\text{``/users/yourself/tmp1''},  \text{``/maple/lib''}$$

```
> with(powers);
```

$$[cube, \; fourth, \; sqr1]$$

The powers package suffers from the minor flaw that you cannot access the commands in the package, not even by long name, unless you load the package using the with command. You can use the readlib command to retrieve the objects stored in a file in a library. The command readlib('name') retrieves all the objects stored in the file name.m in your library. Then readlib returns the object name.

```
> readlib( 'powers' );
```

$$\text{table}([$$

$$fourth = (\mathbf{proc}(x\text{::}anything)$$

$$powers\text{'}_{sqr1}\text{'}\,(powers\text{'}_{sqr1}\text{'}\,(x))$$

$$\mathbf{end}\,)$$

$$sqr1 = (\mathbf{proc}(x\text{::}anything)\, x^2 \; \mathbf{end}\,)$$

$$cube = (\mathbf{proc}(x\text{::}anything)\, x^3 \; \mathbf{end}\,)$$

$$])$$

The fact that readlib('powers') not only defines powers but also returns the table, allows you to access the table in one step.

```
> readlib( 'powers' )[sqr1];
```

$$\mathbf{proc}(x\!::\!anything)\ x^2\ \mathbf{end}$$

```
> readlib( 'powers' )[sqr1](x);
```

$$x^2$$

Thus, instead of storing the table powers in memory, you may define powers as a compact unevaluated call to `readlib`.

```
> powers := 'readlib( 'powers' )';
```

$$powers := \mathrm{readlib}('powers')$$

The first time you use powers, Maple evaluates the `readlib` command which redefines powers as a table of procedures.

```
> powers[fourth](u+v);
```

$$(u + v)^4$$

Each time you start up Maple it reads and executes the commands in a Maple initialization file. On most platforms, the initialization file is called `maple.ini` and it is located in the directory from which you start Maple. To install a package you must place commands similar to the following in the user's initialization file. Using colons, rather than semicolons, suppresses the output.

```
> libname := "/users/yourself/tmp1", libname:
> powers := 'readlib( 'powers' )':
```

With these final touches, you have created your own package. You can now use your package just as you would use any other package in Maple. Packages not only allow you to add special unloaded groups of commands, like those of the `plots` package, when you need them, but they also allow you to package a set of commands so that other users may easily share them.

3.6 Conclusion

The topics in this chapter and chapters 1 and 2 form the building blocks of the programming features in Maple. Although the topics in this chapter are more specialized than those of earlier chapters, they are still very important and are among the most useful. In particular, the first two sections which delve into the workings of procedures which return procedures and local variables are fundamental as you move on to more advanced programming. The later topics, including interactive input, extending Maple, and writing your own packages, while not as fundamental, are also extremely beneficial.

The remaining chapters in this book fall into two categories. Chapters 4 and 5 present in a formal manner the structure of the Maple language and the details of procedures. The other chapters address specific topics, such as plotting, numerical programming, and the Maple debugger.

CHAPTER 4 The Maple Language

This chapter describes the Maple language in detail. The language definition breaks down into four parts: characters, tokens, syntax (how you enter commands), and semantics (the meaning Maple gives to the language). The syntax and semantics are what define a language. Syntax consists of rules to combine words into sentences; syntax is grammar, and is purely mechanical. Semantics is the extra information or meaning that syntax cannot capture, and determines what Maple does when it receives a command.

Syntax   The *syntax* defines what input constitutes a valid Maple expression, statement, or procedure. It answers such questions as:

- Do I need the parentheses in x^(y^z)?
- How do I input a string which is longer than a line?
- How can I input the floating-point number 2.3×10^{-3}?

These are all questions about language *syntax*. They are concerned solely with the input of expressions and programs to Maple, not what Maple does with them.

If the input is not syntactically correct, Maple reports a *syntax* error, and points to where it detected the error. Consider some interactive examples.

Two adjacent minus signs are not valid.

```
> --1;

syntax error, '-' unexpected:
--1;
 ^
```

The character "^" points to where Maple found the error.

Maple accepts many kinds of floating-point formats,

```
> 2.3e-3, 2.3E-03, +0.0023;
```

$$.0023, .0023, .0023$$

but you must place at least one digit between the decimal point and the exponent suffix.

```
> 2.e-3;
```

```
syntax error, missing operator or ';':
2.e-3;              ,
   ^
```

The correct way to write this is 2.0e-3.

Semantics The *semantics* of the language specifies how expressions, statements, and programs execute, that is, what Maple does with them. This answers questions such as:

- Does x/2*z equal x/(2*z) or (x/2)*z? What about x/2/z?
- If x has the value 0, what will happen if I compute $\sin(x)/x$?
- Why does computing $\sin(0)/\sin(0)$ result in 1 and not in an error?
- What is the value of i after executing the following loop?

```
> for i from 1 to 5 do print(i^2) od;
```

The following is a common mistake. Many users think that x/2*z is equal to x/(2*z).

```
> x/2*z, x/(2*z);
```

$$\frac{1}{2}x\,z,\ \frac{1}{2}\frac{x}{z}$$

Syntax Errors in Files Maple reports syntax errors which occur when reading in files and indicates the line number. Write the following program in a file called integrand.

```
f:= proc(x)
       t:= 1 - x^2
       t*sqrt(t)
    end:
```

Then read it in to your Maple session using the read command.

```
> read integrand;

syntax error, missing operator or ';':
        t*sqrt(t)
        ^
```

Maple reports an error at the beginning of line 3. There should be a ";" separating the two calculations, t := 1 - x^2 and t*sqrt(t).

4.1 Language Elements

To simplify the presentation of Maple syntax, consider it in two parts: first, the language *elements* and second, the language *grammar* which explains how to combine the language elements.

The Character Set

The Maple character set consists of letters, digits, and special characters. The letters are the 26 lower-case letters

a, b, c, d, e, f, g, h, i, j, k, l, m, n, o, p, q, r, s, t, u, v, w, x, y, z,

and the 26 upper-case letters

A, B, C, D, E, F, G, H, I, J, K, L, M, N, O, P, Q, R, S, T, U, V, W, X, Y, Z,

The 10 digits are

0, 1, 2, 3, 4, 5, 6, 7, 8, 9.

There are also 32 *special characters*, as shown in Table 4.1. Sections later in this chapter state the uses of each.

Tokens

Maple's language definition combines characters into tokens. Tokens consist of keywords (reserved words), programming-language operators, names, strings, natural integers, and punctuation marks.

Reserved Words   Table 4.2 lists the *reserved words* in Maple. They have special meanings, and thus you cannot use them as variables in programs.

Many other symbols in Maple have predefined meanings. For example, mathematical functions such as sin and cos, Maple commands such as expand and simplify, and type names such as integer and list. However, you can safely use these commands in Maple programs in certain

TABLE 4.1 Special Characters

␣	blank	(	left parenthesis
;	semicolon	)	right parenthesis
:	colon	[	left bracket
+	plus	]	right bracket
−	minus	{	left brace
*	asterisk	}	right brace
/	slash	`	back quote
^	caret	'	single quote (apostrophe)
!	exclamation	"	double quote (ditto)
=	equal	|	vertical bar
<	less than	&	ampersand
>	greater than	_	underscore
@	at sign	%	percent
$	dollar	\\	backslash
.	period	#	sharp
,	comma	?	question mark

contexts. But the reserved words in table 4.2 have a special meanings, and thus you cannot change them.

Programming-Language Operators Three types of *Maple language operators* exist, namely *binary*, *unary*, and *nullary* operators. Tables 4.3 and 4.4 list these operators and their uses. The three nullary operators, %, %%, and %%% are special Maple *names* which refer to the three previously computed expressions. See *The Ditto Operators* on page 155.

Names Maple's language definition predefines many other tokens, including names. For example, mathematical functions like sin and cos, or commands like expand or simplify, or type names like integer or list are all examples of names.

TABLE 4.2 Reserved Words

Keywords	Purpose
if, then, elif, else, fi	
	if statement
for, from, in, by, to,	
while, do, od	for and while loops
proc, local, global, end,	
option,	
options, description	procedures
read, save	read and save statements
quit, done, stop	ending Maple
union, minus, intersect	set operators
and, or, not	Boolean operators
mod	modulus operator

TABLE 4.3   Programming Binary Operators

Operator	Meaning	Operator	Meaning
+	addition	<	less than
−	subtraction	<=	less or equal
*	multiplication	>	greater than
/	division	>=	greater or equal
**	exponentiation	=	equal
^	exponentiation	<>	not equal
$	sequence operator	->	arrow operator
@	composition	union	set union
@@	repeated composition	minus	set difference
&*string*	neutral operator	intersect	set intersection
,	expression separator	::	type declaration
.	string concatenation		pattern binding
.	decimal point	and	logical and
..	ellipsis	or	logical or
mod	modulo	&+	non-commutative
:=	assignment		multiplication

The simplest instance of a *name* consists of letters, digits, and underscores, and does not begin with a number. Maple reserves names beginning with an underscore for internal use only.

Names of the form name are allowed for spreadsheet references. See tutorial 2 in the first chapter of the *Learning Guide* for examples of this usage.

Strings   Maple's language definition also predefines strings. Some simple strings are `"h"`, `"hi"`, `"result"` and `"Input value1"` Generally, enclosing any sequence of characters in double quotes forms a string.

```
> "The modulus should be prime";
```

$$\text{“The modulus should be prime”}$$

TABLE 4.4   Programming Unary Operators

Operator	Meaning
+	unary plus (prefix)
−	unary minus (prefix)
!	factorial (postfix)
$	sequence operator (prefix)
not	logical not (prefix)
&*string*	neutral operator (prefix)
.	decimal point (prefix or postfix)
%*integer*	label (prefix)

```
> "There were %d values";
```

$$\text{``There were \%d values''}$$

You should not confuse the double quote character, ", which delimits a string, with the back quote character, ', which forms a *symbol* or the single quote, ', which delays evaluation. A string's length has no practical limit in Maple. On most Maple implementations, this means that a string can contain more than half a million characters.

To make the double quote character appear in a string, type a backslash character and a double quote (") where you want the double quote character to appear.

```
> "a\"b";
```

$$\text{"a\"b"}$$

Similarly, to allow a backslash (escape character) to appear as one of the characters in a string, type two consecutive backslashes \\.

```
> "a\\b";
```

$$\text{"a\\b"}$$

The special backslash characters mentioned above only count as one character, as is demonstrated by using the length command.

```
> length(%);
```

$$3$$

A reserved word enclosed in double quotes also becomes a valid Maple string, distinct from its usage as a token.

```
> "while";
```

$$\text{``while''}$$

The enclosing double quotes themselves do not form part of the string.

```
> length("abcde");
```

$$5$$

To access individual characters or substrings, strings can be subscripted in much the same way as lists. An integer range provides access.

```
> S := "This is a string";
```

$$S := \text{``This is a string''}$$

```
> S[6..9];
```

$$\text{``is a''}$$

```
> S[-6..-1];
```

$$\text{“string”}$$

As well, iterations can be performed over the characters in a string.

```
> seq(i,i="over a string");
```

$$\text{“o”, “v”, “e”, “r”, “ ”, “a”, “ ”, “s”, “t”, “r”, “i”, “n”, “g”}$$

Integers   A *natural integer* is any sequence of one or more digits. Maple ignores any leading zeroes.

```
> 0314159265358979323846264З;
```

$$3141592653589793238462643$$

The length limit for integers is system-dependent, but is generally much larger than users require. For example, the length limit on most 32-bit computers is 524 280 decimal digits.

An *integer* is either a natural integer or a signed integer. Either *+natural* or *−natural* indicates a signed integer.

```
> -12345678901234567890;
```

$$-12345678901234567890$$

```
> +12345678901234567890;
```

$$12345678901234567890$$

Token Separators

You can separate tokens using either *white space* or punctuation marks. This tells Maple where one token ends and the next begins.

Blanks, Lines, Comments, and Continuation   The *white space* characters are space, tab, return, and line-feed. This book uses the terminology *new-line* to refer to either return or line-feed since the Maple system does not distinguish between these characters. The terminology *blank* refers to either space or tab. The white space characters separate tokens, but are not themselves tokens.

White space characters cannot normally occur within a token.

```
> a: = b;
```

```
syntax error, ‘=‘ unexpected:
a: = b;
      ^
```

You can use white space characters freely between tokens.

```
> a * x + x*y;
```

$$a\,x + x\,y$$

The only place where white space can become part of a token is in a string formed by enclosing a sequence of characters within back quotes, in which case the white space characters are as significant as any other character.

On a line, unless you are in the middle of a string, Maple considers all characters which follow a sharp character "#" to be part of a *comment*.

Since white space and newline characters are functionally the same, you can continue statements from line to line.

```
> a:= 1 + x +
>     x^2;
```

$$a := 1 + x + x^2$$

The problem of continuation from one line to the next is less trivial when long numbers or long strings are involved since these two classes of tokens are not restricted to a few characters in length. The general mechanism in Maple to specify continuation of one line onto the next is as follows: if the special character backslash, \, immediately precedes a newline character, then the parser ignores both the backslash and the newline. If a backslash occurs in the middle of a line, Maple usually ignores it; see ?backslash for exceptions. You can use this to break up a long sequence of digits into groups of smaller sequences, to enhance readability.

```
> "The input should be either a list of\
>  variables or a set of variables";
```

$$\text{``The input should be either a list of variables or } \backslash$$
$$\text{a set of variables''}$$

```
> G:= 0.5772156649\0153286060\
> 6512090082\4024310421\5933593992;
```

$$G := .5772156649015328606065120900824024\backslash$$
$$3104215933593992$$

Punctuation Marks Table 4.5 lists the *punctuation marks*.

; and : Use the semicolon and the colon to separate statements. The distinction between these marks is that a colon during an interactive session prevents the result of the statement from printing.

TABLE 4.5 Maple Punctuation Marks

;	semicolon	(	left parenthesis
:	colon	)	right parenthesis
'	single quote	[	left bracket
`	back quote	]	right bracket
|	vertical bar	{	left brace
<	left angle bracket	}	right brace
>	right angle bracket	,	comma

```
> f:=x->x^2; p:=plot(f(x), x=0..10):
```

' Enclosing an expression, or part of an expression, in a pair of single quotes delays evaluation of the expression (subexpression) by one level. See *Unevaluated Expressions* on page 167.

```
> ''sin''(Pi);
```

$$'\text{sin}'(\pi)$$

```
> %;
```

$$\sin(\pi)$$

```
> %;
```

$$0$$

` To form symbols, use the back quote character.

```
> limit(f(x), x=0, `right`);
```

() The left and right parentheses group terms in an expression and group parameters in a function call.

```
> (a+b)*c; cos(Pi);
```

[] Use the left and right square brackets to form indexed (subscripted) names and to select components from aggregate objects such as arrays, tables, and lists. See *Selection Operation* on page 146.

```
> a[1]; L:=[2,3,5,7]; L[3];
```

[] and {} Use the left and right square brackets also to form lists, and the left and right braces to form sets. See *Sets and Lists* on page 145.

```
> L:=[2,3,5,2]; S:={2,3,5,2};
```

<> The left and right angle brackets form a grouping which you may define.

```
> <2,3,5,7>;
```

, Use the comma to form a sequence, to separate the arguments of a function call, and to separate the elements of a list or set. See *Sequences* on page 142.

```
> sin(Pi), 0, limit(cos(xi)/xi, xi=infinity);
```

4.2 Escape Characters

The *escape characters* are ?, !, #, and \. Their special meanings are outlined below.

? The question mark character, if it appears as the first non-blank character on a line, invokes Maple's *help* facility. The words following ? on the same line determine the arguments to the help procedure. Use either "," or "/" to separate the words.

! The exclamation mark character, if it appears as the first non-blank character on a line, passes the remainder of the line as a command to the host operating system. This facility is not available on all platforms.

The hash mark character indicates that Maple is to treat the characters following it on the line as a *comment*. In other words, Maple ignores them. They have no effect on any calculation that Maple might do.

\ Use the backslash character for *continuation* of lines and for grouping of characters within a token. See *Blanks, Lines, Comments, and Continuation* on page 113.

4.3 Statements

There are eight types of statements in Maple. They are the

1. assignment statement
2. selection statement
3. repetition statement
4. read statement
5. save statement
6. empty statement
7. quit statement
8. expressions

Expressions on page 131 discusses expressions at length.

Throughout the remainder of this section, *expr* stands for any expression, *statseq* stands for a sequence of statements separated by semicolons.

The Assignment Statement

The syntax of the assignment statement is

```
name := expr;
name_1, ..., name_n := expr_1, ..., expr_n;
```

This assigns, or sets, the value of the variable *name* to be the result of executing the expression *expr*. Multiple assignments can also be performed.

Names A *name* in Maple may be a *symbol* or an *indexed name* (subscript). Names stand for unknowns in formulae. They also serve as programming variables. A name only becomes a programming variable when Maple assigns it a value. Otherwise, if Maple does not assign the name a value, then it remains an unknown.

```
> 2*y - 1;
```

$$2y - 1$$

```
> x := 3; x^2 + 1;
```

$$x := 3$$

$$10$$

```
> a[1]^2;   a[1] := 3;   a[1]^2;
```

$$a_1{}^2$$

$$a_1 := 3$$

$$9$$

```
> f[Cu] := 1.512;
```

$$f_{Cu} := 1.512$$

To define a function, use the *arrow notation*, ->.

```
> phi := t -> t^2;
```

$$\phi := t \to t^2$$

Note that the following does *not* define a function; instead an entry is created in the remember table for phi. See *Remember Tables* on page 68.

```
> phi(t) := t^2;
```

$$\phi(t) := t^2$$

Mapping Notation on page 182 contains more on how to define functions.

Indexed Names   Another form of a name in Maple is the *indexed name* or subscripted name, which has the form

> | name [sequence] |

Note that since an indexed name is itself a valid name, you can add a succession of subscripts.

```
> A[1,2];
```

$$A_{1,2}$$

```
> A[i,3*j-1];
```

$$A_{i,3j-1}$$

```
> b[1][1], data[Cu,gold][1];
```

$$b_{1\,1}, \; data_{Cu,gold_1}$$

The use of the indexed name A[1,2] does not imply that A is an array, as in some languages. The statement

```
> a := A[1,2] + A[2,1] - A[1,1]*A[2,2];
```

$$a := A_{1,2} + A_{2,1} - A_{1,1}\,A_{2,2}$$

forms a formula in the four indexed names. (However, if A does evaluate to an array or table, then A[1,1] refers to the (1, 1) element of the array or table.)

The Concatenation Operator   Generally, you can form a *name* using the *concatenation* operator in one of the following three forms.

> | name . natural |
> | name . string |
> | name . (expression) |

Since a *name* can appear on the left-hand side, Maple allows a succession of concatenations. Some examples of the use of the concatenation operator for name formation are given in Table 4.6.

The concatenation operator is a binary operator which requires a name or a string as its left operand. Although Maple usually evaluates expressions

TABLE 4.6   Maple Concatenation Operator

v.5	p."n"	a.(2*i)	V.(1..n)	r.i.j

from left to right, it evaluates concatenations from right to left. Maple evaluates the right-most operand, then concatenates to the left operand. If it evaluates the right operand to an integer, string or name, then the result of the concatenation is a string or name (depending on the type of the left-most operand). If it evaluates the right operand to some other type of object, say a formula, then the result of the operation is an unevaluated concatenated object.

```
> p.n;
```

$$pn$$

```
> "p".n;
```

$$``pn"$$

```
> n := 4: p.n;
```

$$p4$$

```
> p.(2*n+1);
```

$$p9$$

```
> p.(2*m+1);
```

$$p.(2\,m + 1)$$

If the right hand *expression* is a sequence or a range and the operands of the range are integers or character strings, then Maple creates a sequence of names.

```
> x.(a, b, 4, 67);
```

$$xa,\ xb,\ x4,\ x67$$

```
> x.(1..5);
```

$$x1,\ x2,\ x3,\ x4,\ x5$$

```
> X.("a".."g");
```

$$Xa,\ Xb,\ Xc,\ Xd,\ Xe,\ Xf,\ Xg$$

If more than one range appears, it composes the extended sequence of names.

```
> x.(1..2).(1..3);
```

$$x11, x12, x13, x21, x22, x23$$

Maple never fully evaluates the left-most object, but rather evaluates it to a name. Concatenations can also be formed with the cat command.

cat(*sequence*)

Note that all the arguments of the cat command are evaluated normally (as for any other function call); therefore

```
> cat( "a", "b", "c" );
```

"abc"

is equivalent to

```
> "" . a . b . c;
```

"abc"

Protected Names   Many names in Maple have a predefined meaning, and you cannot directly assign a value to them. For example, the names of built-in functions such as sin, the sine function, utility operations such as degree, which computes the degree of a polynomial, commands such as diff for differentiation, and type names like integer and list, are all protected names. When the user attempts to assign to any of these names, an error occurs.

```
> list := [1,2];

Error,
attempting to assign to 'list' which is protected
```

The system protects these names from accidental assignment. It *is* possible to assign to these names by first unprotecting them as follows.

```
> unprotect(sin);
> sin := "a sin indeed";
```

sin := "a sin indeed"

However, now the areas of Maple that rely on the sine function will not work properly.

```
> plot( 1, 0..2*Pi, coords=polar );

Warning in iris-plot: empty plot
```

On the other hand, if you write programs in which you want to prevent a user from assigning to certain names, do so using the protect command.

```
> mysqr := x -> x^2;
```

$$mysqr := x \rightarrow x^2$$

```
> protect( mysqr );
> mysqr := 9;
```

```
Error,
attempting to assign to 'mysqr' which is protected
```

Unassignment: Clearing a Name

When names do not carry assigned values they act as unknowns. When assigned values, they act as variables. It is often desirable to *unassign* (or clear) a name which previously carried an assigned value, so that you can use the name as an unknown again. The way to do this in Maple is to *assign the name to be itself.* Maple understands this to mean clear the name. The command

$$\boxed{\texttt{evaln(\textit{name})}}$$

evaluates *name* to a name (as opposed to evaluating *name* to its value as in other function calls). You can thus unassign a name as follows.

```
> a := evaln(a);
```

$$a := a$$

```
> i := 4;
```

$$i := 4$$

```
> a[i] := evaln(a[i]);
```

$$a_4 := a_4$$

```
> a.i := evaln(a.i);
```

$$a4 := a4$$

In the special case where *name* is a string you may also unassign a variable by delaying evaluation of the right-hand side with single quotes (`'`). See *Unevaluated Expressions* on page 167.

```
> a := 'a';
```

$$a := a$$

Related Functions   You can use the `assigned` command to test if a name has an assigned value.

```
> assigned(a);
```

$$false$$

The `assign` command assigns a variable.

```
> assign( a=b );
> assigned(a);
```

$$true$$

```
> a;
```

$$b$$

Maple normally evaluates all the arguments of `assign`. Therefore, because of the previous assignment, `assign(a=b)`, Maple assigns b the value 2 here.

```
> assign( a=2 );
> b;
```

$$2$$

One level evaluation of a reveals that a still has the value b.

```
> eval( a, 1 );
```

$$b$$

Changing the value of a does not affect the value of b.

```
> a := 3;
```

$$a := 3$$

```
> b;
```

$$2$$

Often, applications of the `assign` command are to a set or list of equations.

```
> eqn1   :=  x + y = 2:
> eqn2   :=  x - y = 3:
> sol := solve( {eqn1, eqn2}, {x, y} );
```

$$sol := \left\{ y = \frac{-1}{2}, \; x = \frac{5}{2} \right\}$$

Maple assigns the variables x and y according to the set sol of equations.

```
> assign(sol);
```

```
> x;
```

$$\frac{5}{2}$$

```
> assigned(x);
```

true

It is recommended that you not assign values to expressions like f(x). See *Remember Tables* on page 68 for details.

The Selection Statement

The selection or conditional statement has four forms. The syntax of the first two forms is

```
if expr then statseq fi;
if expr then statseq1 else statseq2 fi;
```

Maple executes the selection statement as follows. It evaluates the expression in the if clause (*expr*). If the result is the boolean value true, then Maple executes the statement sequence in the then clause. If the result is the boolean value false or FAIL, then Maple executes the statements in the else clause.

```
> x := -2:
> if x<0 then 0 else 1 fi;
```

$$0$$

The *expr* must evaluate to one of the boolean values true, false, or FAIL; see *Boolean Expressions* on page 161.

```
> if x then 0 else 1 fi;

Error, invalid boolean expression
```

Omit the else clause if you do not want to include an alternative course of action when the condition is false.

```
> if x>0 then x := x-1 fi;
> x;
```

$$-2$$

The selection statement may be nested, that is, the statement sequence in the then clause or else clause may be any statement, including an if statement.

Compute the sign of a number.

```
> if x > 1 then 1
> else if x=0 then 0 else -1 fi
> fi;
```

The following example demonstrates a use of FAIL.

```
> r := FAIL:
> if r then
>     print(1)
> else
>     if not r then
>         print(0)
>     else
>         print(-1)
>     fi
> fi;
```

$$-1$$

If Maple has many cases to consider, the use of nested if statements becomes messy and unreadable. Maple provides the following two alternatives.

```
if expr then statseq elif expr then statseq fi;
if expr then statseq elif expr then statseq
else statseq fi;
```

The elif expr then statseq construct may appear more than once.

Here you can implement the sign function using an elif clause

```
> x := -2;
```

$$x := -2$$

```
> if x<0 then -1
> elif x=0 then 0
> else 1
> fi;
```

$$-1$$

In this form, you can view the selection statement as a case statement with the optional else clause as the default case. For example, if you are writing a program that accepts a parameter n with four possible values, $0, 1, 2, 3$, then you might write

```
> n := 5;
```

$$n := 5$$

```
> if   n=0 then 0
> elif n=1 then 1/2
> elif n=2 then sqrt(2)/2
> elif n=3 then sqrt(3)/2
> else ERROR("bad argument", n)
> fi;
```

```
Error, bad argument, 5
```

The Repetition Statement

The most general repetition statement in Maple is the for loop. However, you can replace many loops with more efficient and concise special forms. See *Useful Looping Constructs* on page 172.

The for loop has two forms: the for-from loop and the for-in loop.

The for-from Loop   A typical for-from loop has the form

```
> for i from 2 to 5 do i^2 od;
```

$$4$$

$$9$$

$$16$$

$$25$$

This sequence of results arose as follows. First, Maple assigns i the value 2. Since 2 is less than 5, Maple executes the statement between the do and the od. Then it increments i by 1 to 3, tests again, the loop executes, and so on until i is (strictly) larger than 5. In this case the final value of i is

```
> i;
```

$$6$$

The syntax of the for-from loop is

```
for name from expr by expr to expr while expr
do statseq od;
```

You may omit any of the clauses for *name*, from *expr*, by *expr*, to *expr*, or while *expr*. You may omit the sequence of statements *statseq*. Except for the for clause, which must always appear first, the other clauses may appear in any order. If you omit a clause, it has a default value, which is shown in table 4.7.

TABLE 4.7 Clauses and Their Default Values

Clause	Default value
for	dummy variable
from	1
by	1
to	infinity
while	true

You could also write the example above as

```
> for i from 2 by 1 to 5 while true do i^2 od:
```

If the by clause is negative, the for loop counts downward.

```
> for i from 5 to 2 by -1 do i^2 od;
```

$$25$$

$$16$$

$$9$$

$$4$$

To find the first prime number greater than 10^7 you could write

```
> for i from 10^7 while not isprime(i) do od;
```

Now i is the first prime larger than 10^7.

```
> i;
```

$$10000019$$

Notice that the body of the loop is empty. Maple allows for the empty statement. Try improving the program by considering only the odd numbers.

```
> for i from 10^7+1 by 2 while not isprime(i) do od;
> i;
```

$$10000019$$

Here is an example of repeating an action n times. Throw a die five times

```
> die := rand(1..6):
> to 5 do die(); od;
```

$$4$$

$$3$$

$$4$$

$$6$$

$$5$$

Omitting all clauses produces an infinite loop.

```
do statseq od;
```

This is equivalent to

```
for name from 1 by 1 to infinity while true
do statseq od;
```

Such a loop will loop forever unless the break construct (see *The* break *and* next *Commands* on page 128) or a procedure RETURN (see *Explicit Returns* on page 197) terminates it, or if Maple encounters the quit statement, or if an error occurs.

The while Loop   The while loop is a for loop with all its clauses omitted except the while clause, that is

```
while expr do statseq od;
```

The expression *expr* is called the *while condition*. It must be a boolean valued expression, that is, it must evaluate to true, false, or FAIL. For example

```
> x := 256;
```

$$x := 256$$

```
> while x>1 do x := x/4 od;
```

$$x := 64$$
$$x := 16$$
$$x := 4$$
$$x := 1$$

The while loop works as follows. First, Maple evaluates the while condition. If it evaluates to true, Maple executes the body of the loop. This loop repeats until the while condition evaluates to false or FAIL. Note, Maple evaluates the while condition *before* it executes the body of the loop. An error occurs if the while condition does not evaluate to one of true, false, or FAIL.

```
> x := 1/2:
> while x>1 do x := x/2 od;
```

```
> x;
```

$$\frac{1}{2}$$

```
> while x do x := x/2 od;
Error, invalid boolean expression
```

The for-in Loop   Suppose you have a list of integers L and want to find the integers in the list that are at most 7. You could write

```
> L := [7,2,5,8,7,9];
```

$$L := [7, 2, 5, 8, 7, 9]$$

```
> for i in L do
>     if i <= 7 then print(i) fi;
> od;
```

$$7$$

$$2$$

$$5$$

$$7$$

This example cycles through the components of an object. The object, this time, is a list. But in other examples, the object might be a set, a sum of terms, a product of factors, or the characters of a string. The syntax for the for-in loop is

> for *name* in *expr* while *expr* do *statseq* od;

The loop index (the *name* specified in the for clause of the statement) takes on the operands of the first *expr*. See *Expression Trees: Internal Representation* on page 131 for a description of the operands associated with each data type. You can test the value of the index in the optional while clause, and, of course, the value of the index is available when you execute the *statseq*. Note, the value of the index variable name remains assigned at the end of the loop if the object contains at least one operand.

The break and next Commands   Within the Maple language reside two additional loop control constructs: break and next. When Maple evaluates the special name break, the result is to exit from the innermost repetition statement within which it occurs. Execution then proceeds with the first statement following this repetition statement.

```
> L := [2, 5, 7, 8, 9];
```

$$L := [2, 5, 7, 8, 9]$$

```
> for i in L do
>      print(i);
>      if i=7 then break fi;
> od;
```

$$2$$
$$5$$
$$7$$

When Maple evaluates the special name next, it then proceeds immediately to the next iteration. For example, suppose you want to skip over the elements in a list that are equal to 7.

```
> L := [7,2,5,8,7,9];
```

$$L := [7, 2, 5, 8, 7, 9]$$

```
> for i in L do
>      if i=7 then next fi;
>      print(i);
> od;
```

$$2$$
$$5$$
$$8$$
$$9$$

An error occurs if Maple evaluates the names break or next in a context other than within a repetition statement.

```
> next;
```

```
Error, break or next not in loop
```

The read and save Statements

The file system is an important part of Maple. The user may interact with the file system either explicitly by using the read and save statements, or implicitly by executing a command that automatically loads information from a file. For example, the computation of an integral may load many commands from the Maple library. The read and save statements read and save Maple data and programs to and from files. See also Chapter 9.

Saving a Maple Session   The save statement allows you to save the values of a sequence of variables. It takes the general form

```
save nameseq, filename;
```

Here *nameseq* must be a sequence of names of assigned variables. Maple saves each variable name and its value in an assignment statement in the file *filename*. Maple evaluates each argument, except the last one, to a name. It evaluates the last argument normally.

Clear Maple using the `restart` command and assign three new values.

```
> restart:
> r0 := x^3:
> r1 := diff(r0,x):
> r2 := diff(r1,x):
```

The next statement saves r0, r1 and r2 in the ASCII file `my_file`:

```
> save r0, r1, r2, "my_file";
```

This is now the contents of the file `my_file`.

```
r0 := x^3;
r1 := 3*x^2;
r2 := 6*x;
```

The expression *filename* must evaluate to a name which specifies the name of a file. If the name ends with the characters ".m" then Maple saves the variables and their values in its internal format. Internal format files are not intended for humans to read; Maple encodes them in a binary format. They are compact, and are faster for Maple to read than text files. If the file name does not end in ".m" then Maple writes the file in ASCII text format.

A special form of the `save` statement saves the values of all the assigned variables in the current Maple session. This special "session save" feature is available only for saves into ".m" files using the internal save format. To save the entire session, no variable names are listed and the file name *must* end in ".m."

The next statement saves all of the variables in the current session in an internal file format, in the file `rvalues.m`.

```
> save "rvalues.m";
```

You can read it back into Maple later using the `read` command.

The `read` Statement The read statement

> read *filename*;

reads a file into the Maple session. The *filename* must evaluate to the name of the file. The file must be either a Maple internal format file (a .m file), or a text file.

Reading Text Files If the file is a plain text file, then it must contain a sequence of valid Maple statements, separated by semicolons or colons. The effect of reading the file is identical to entering the same sequence of statements interactively. The system displays the result of executing each statement that it reads in from the file.

4.4 Expressions

Expressions are the fundamental entities in the Maple language. The various types of expressions include constants, names of variables and unknowns, formulae, boolean expressions, series, and other data structures. Technically speaking, procedures are also valid expressions since you may use them wherever an expression is legal. Chapter 5 describes them separately. The ?precedence help page gives the order of precedence of all programming-language operators.

Expression Trees: Internal Representation
Consider the following formula.

```
> f := sin(x) + 2*cos(x)^2*sin(x) + 3;
```

$$f := \sin(x) + 2\cos(x)^2 \sin(x) + 3$$

To represent this formula, Maple builds an *expression tree*.

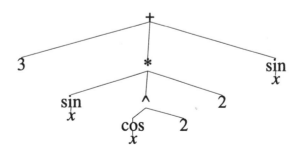

The first node of the expression tree labeled "+" is a sum. This indicates the expression's *type*. This expression has three branches corresponding to the three terms in the sum. The nodes of each branch tell you each term's type in the sum. And so on down the tree until you get to the leaves of the tree, which are names and integers in this example.

TABLE 4.8   Primitive Functions

type(*f*, *t*)	tests if *f* is of type *t*
nops(*f*)	returns the number of operands of *f*
op(*i*, *f*)	selects the *i*th operand of *f*
subsop(*i=g*, *f*)	replaces the *i*th operand of *f* with *g*

When programming with expressions, you need a way to determine what type of expression you have, how many operands or branches an expression has, and a way of selecting those operands. You also need a way of building new expressions, for example, by replacing one operand of an expression with a new value. Table 4.8 lists the primitive functions for doing this.

```
> type(f, '+');
```

$$true$$

```
> type(f, '*');
```

$$false$$

```
> nops(f);
```

$$3$$

```
> op(1, f);
```

$$\sin(x)$$

```
> subsop(2=0, f);
```

$$\sin(x) + 3$$

By determining the type of an expression, the number of operands it has, and selecting each operand of the expression, you can systematically work all the way through an expression.

```
> t := op(2, f);
```

$$t := 2\cos(x)^2\sin(x)$$

```
> type(t, '*');
```

$$true$$

```
> nops(t);
```

$$3$$

```
> type(op(1,t), integer);
```

$$true$$

```
> type(op(2,t), '^');
```

$$true$$

```
> type(op(3,t), function);
```

$$true$$

The op command has several other useful forms. The first is

$$\boxed{op(i..j, \; f)}$$

which returns the sequence

$$\boxed{op(i, \; f), \; op(i+1, \; f), \; ..., \; op(j-1, \; f), \; op(j, \; f)}$$

of operands of f. You may want to see the whole sequence of operands of an expression. You can do this with

$$\boxed{op(f)}$$

which is equivalent to op(1..nops(f),f). The special operand op(0,f) generally returns the type of an expression. An exception occurs when f is a function, in which case it returns the name of the function.

```
> op(0, f);
```

$$`+`$$

```
> op(1..3, f);
```

$$\sin(x), \; 2\cos(x)^2 \sin(x), \; 3$$

```
> op(0, op(1,f));
```

$$\sin$$

```
> op(0, op(2,f));
```

$$`*`$$

```
> op(0, op(3,f));
```

$$integer$$

Evaluation and Simplification   Consider this example in detail.

```
> x := Pi/6:
> sin(x) + 2*cos(x)^2*sin(x) + 3;
```

$$\frac{17}{4}$$

What does Maple do when it executes the second command? Maple first reads and *parses* the input line. As it is parsing the input line it builds an expression tree to represent the value

$$\sin(x) + 2\cos(x)^2\sin(x) + 3.$$

Next it *evaluates* the expression tree, then *simplifies* the result. Evaluation means substituting values for variables and invoking any functions. Here x evaluates to $\pi/6$, hence with these substitutions the expression is as follows.

$$\sin(\pi/6) + 2\cos(\pi/6)^2\sin(\pi/6) + 3$$

Invoking the sin and cos functions, Maple obtains a new expression tree,

$$1/2 + 2 \times (1/2\sqrt{3})^2 \times 1/2 + 3.$$

Finally Maple simplifies this expression tree (does the arithmetic) to obtain the fraction 17/4. In the following example, evaluation occurs, but no simplification is possible.

```
> x := 1;
```

$$x := 1$$

```
> sin(x) + 2*cos(x)^2*sin(x) + 3;
```

$$\sin(1) + 2\cos(1)^2\sin(1) + 3$$

We now present in detail every kind of expression, beginning with the constants. The presentation states how to input the expression, gives examples of how and where to use the expression, and the action of the type, nops, op, and subsop commands on the expression.

The numeric constants in Maple are integers, fractions, and floating-point or decimal numbers. The complex numeric constants are the complex integers (Gaussian integers), complex rationals, and complex floating-point numbers.

TABLE 4.9   Subtypes of Integers

negint	negative integer
posint	positive integer
nonnegint	non-negative integer
even	even integer
odd	odd integer
prime	prime integer

The Types and Operands of Integers, Strings, Indexed Names, and Concatenations

The type of an integer is integer. The type command also understands the subtypes of integers listed in table 4.9. The op and nops commands consider an integer to have only one operand, namely, the integer itself.

```
> x := 23;
```

$$x := 23$$

```
> op(0, x);
```

integer

```
> op(x);
```

23

```
> type(x, prime);
```

true

The type of a string is string. A string also has only one operand; the string itself.

```
> s := "Is this a string?";
```

$$s := \text{“Is this a string?”}$$

```
> type(s, string);
```

true

```
> nops(s);
```

1

```
> op(s);
```

"Is this a string?"

The type of an indexed name is indexed. The operands of an indexed name are the indices or subscripts and the zeroth operand is the base name. The type command also understands the composite type name which Maple defines as either a string or an indexed name.

```
> x := A[1] [2,3];
```

$$x := A_{12,3}$$

```
> type(x, indexed);
```

true

```
> nops(x);
```

2

```
> op(x);
```

2, 3

```
> op(0,x);
```

$$A_1$$

```
> y:=%;
```

$$y := A_1$$

```
> type(y, indexed);
```

true

```
> nops(y), op(0,y), op(y);
```

$$1, A, 1$$

The type of an unevaluated concatenation is ".". This type has two operands, the left-hand side expression and the right-hand side expression.

```
> c := p.(2*m + 1);
```

$$c := p.(2\,m + 1)$$

```
> type(c, '.');
```

true

```
> op(0, c);
```

..

```
> nops(c);
```

2

```
> op(c);
```

$$p,\ 2m+1$$

Fractions and Rational Numbers

A *fraction* is input as

$$\boxed{integer/natural}$$

Maple does arithmetic with fractions and integers *exactly*. Maple always immediately simplifies a fraction so that the denominator is positive, and reduces the fraction to lowest terms by canceling out the greatest common divisor of the numerator and denominator.

```
> -30/12;
```

$$\frac{-5}{2}$$

If the denominator is 1 after simplification of a fraction, Maple automatically converts it to an integer. The type of a fraction is `fraction`. The type command also understands the composite type name `rational`, which is an `integer` or a `fraction`, that is, a rational number.

```
> x := 4/6;
```

$$x := \frac{2}{3}$$

```
> type(x,rational);
```

$$true$$

A fraction has two operands, the numerator and denominator. In addition to the op command, you may use the commands `numer` and `denom` to extract the numerator and denominator of a fraction, respectively.

```
> op(1,x), op(2,x);
```

$$2,\ 3$$

```
> numer(x), denom(x);
```

$$2,\ 3$$

Floating-Point (Decimal) Numbers

An *unsigned float* has one of the following six forms:

> natural.natural
> natural.
> .natural
> natural exponent
> natural.natural exponent
> .natural exponent

where the *exponent* suffix is the letter "e" or "E" followed by a signed integer with no spaces in the middle. A *floating-point number* is an *unsigned_float* or a signed float (*+unsigned_float* or *–unsigned_float* indicates a signed float).

```
> 1.2,   -2., +.2;
```

$$1.2, -2., .2$$

```
> 2e2,  1.2E+2, -.2e-2;
```

$$200., 120., -.002$$

Note that

```
> 1.e2;

syntax error, missing operator or ';':
1.e2;
    ^
```

is not valid, and that spaces are significant.

```
> .2e -1 <> .2e-1;
```

$$-.8 \neq .02$$

The type of a floating-point number is `float`. The `type` command also understands composite type name `numeric` which Maple defines to be an `integer`, `fraction`, or `float`.

A floating-point number has two operands, the significand m and the exponent e; that is, the number represented is $m \times 10^e$.

```
> x := 231.3;
```

$$x := 231.3$$

```
> op(1,x);
```

$$2313$$

```
> op(2,x);
```

$$-1$$

An alternative input format for floating-point numbers in Maple is to use the `Float` command

$$\boxed{\text{Float}(m, \ e)}$$

which creates the floating-point number $m \times 10^e$ from the significand m and the exponent e.

Maple represents the significand of a floating-point number as an integer. The length restriction is typically greater than 500 000 digits of precision. However, Maple restricts the exponent to a machine or word-size integer. This size is system-dependent, but is typically nine or more digits in length. You can also input a floating-point number $m \times 10^e$ by multiplying by a power of 10. But Maple then computes the expression 10^e before multiplying by the significand. This method is inefficient for large exponents.

Arithmetic with Floating-Point Numbers   For arithmetic operations and the standard mathematical functions, if one of the operands (or arguments) is a floating-point number or evaluates to a floating-point number, then floating-point arithmetic takes place automatically. The global name `Digits`, which has 10 as its default, determines the number of digits which Maple uses when calculating with floating-point numbers (the number of digits in the significand).

```
> x := 2.3:   y := 3.7:
> 1 - x/y;
```

$$.3783783784$$

In general, you may use the `evalf` command to force the evaluation of a non-floating-point expression to a floating-point expression where possible.

```
> x := ln(2);
```

$$x := \ln(2)$$

```
> evalf(x);
```

$$.6931471806$$

An optional second argument to the `evalf` command specifies the precision at which Maple is to do this evaluation.

```
> evalf(x,15);
```

$$.693147180559945$$

TABLE 4.10   Types of Complex Numbers

`complex(integer)`	both *a* and *b* are integers, possibly 0
`complex(rational)`	both *a* and *b* are rationals
`complex(float)`	both *a* and *b* are floating-point constants
`complex(numeric)`	any of the above

Complex Numerical Constants

By default, I denotes the complex unit $\sqrt{-1}$ in Maple. In fact, all of the following are equivalent.

```
> sqrt(-1), I, (-1)^(1/2);
```

$$I, I, I$$

A complex number $a+bi$ is input as the sum a + b*I in Maple; that is, Maple does not utilize special representations for complex numbers. Use the commands Re and Im to select the real and imaginary parts, respectively.

```
> x := 2+3*I;
```

$$x := 2 + 3\,I$$

```
> Re(x), Im(x);
```

$$2, 3$$

The type of a complex number is `complex(numeric)`. This means that the real and imaginary parts are of type `numeric`, that is, integers, fractions, or floating-point numbers. Other useful type names are listed in table 4.10.

Arithmetic with complex numbers is done automatically.

```
> x := (1 + I);   y := 2.0 - I;
```

$$x := 1 + I$$
$$y := 2.0 - 1.\,I$$

```
> x+y;
```

$$3.0$$

Maple also knows how to evaluate elementary functions and many special functions over the complex numbers. It does this automatically if *a* and *b* are numeric constants and one of *a* or *b* is a decimal number.

```
> exp(2+3*I), exp(2+3.0*I);
```

$$e^{(2+3\,I)}, -7.315110095 + 1.042743656\,I$$

If the arguments are not complex floating-point constants, you can expand the expression in some cases into the form $a + bi$, where *a* and *b* are real, using the `evalc` command.

Here the result is not in the form $a + bi$ since a is not of type `numeric`

```
> 1/(a - I);
```

$$\frac{1}{a - I}$$

```
> evalc(%);
```

$$\frac{a}{a^2 + 1} + \frac{I}{a^2 + 1}$$

If you prefer to use another letter, say j, for the imaginary unit, use the `alias` command as follows.

```
> alias( I=I, j=sqrt(-1) );
```

$$j$$

```
> solve( {z^2=-1}, {z} );
```

$$\{z = j\}, \{z = -j\}$$

The following command removes the alias for j and reinstates I as the imaginary unit.

```
> alias( I=sqrt(-1), j=j );
```

$$I$$

```
> solve( {z^2=-1}, {z} );
```

$$\{z = I\}, \{z = -I\}$$

Labels

A *label* in Maple has the form

%natural

that is, the unary operator % followed by a natural integer. The percentage sign takes on double duty, as a label and as the ditto operator, which represents the result of the last one, two, or three commands.

A label is only valid after Maple's pretty-printer introduces it. The purpose is to allow the naming (labeling) of common subexpressions, which serves to decrease the size of the printed output, making it more comprehensible. After the pretty-printer introduces it, you may use a label just like an assigned name in Maple.

```
> solve( {x^3-y^3=2, x^2+y^2=1}, {x, y} );
```

$$\{y = \%1, \ x = -\frac{1}{3}\%1\,(-4\,\%1^3 + 2\,\%1^4 + 6\,\%1 - 3 - \%1^2)$$

$$\}$$
$$\%1 := \mathrm{RootOf}(3\,\_Z^2 + 3 - 3\,\_Z^4 + 2\,\_Z^6 + 4\,\_Z^3)$$

After you obtain the above printout, the label %1 is an assigned name and its value is the RootOf expression shown.

```
> %1;
```

$$\mathrm{RootOf}(3\,\_Z^2 + 3 - 3\,\_Z^4 + 2\,\_Z^6 + 4\,\_Z^3)$$

Two options are available for adjusting this facility. The option

> interface(labelwidth=*n*)

specifies that Maple should not display expressions less than *n* characters wide (approximately) as labels. The default is 20 characters. You may turn off this facility entirely using

```
> interface(labelling=false);
```

Sequences

A *sequence* is an expression of the form

> expression_1, expression_2, ..., expression_n

The comma operator joins expressions into a sequence. It has the lowest precedence of all operators except assignment. A key property of sequences is that if any of *expression_i* themselves are sequences, this flattens out the result into a single unnested sequence.

```
> a := A, B, C;
```

$$a := A, \ B, \ C$$

```
> a,b,a;
```

$$A, \ B, \ C, \ b, \ A, \ B, \ C$$

A zero-length sequence is syntactically valid. It arises, for example, in the context of forming an empty list, an empty set, a function call with no parameters, or an indexed name with no subscripts. Maple initially assigns the special name NULL to the zero-length sequence, and you may use it whenever necessary.

You cannot use the type command to test the type of a sequence, nor can you use the nops or op commands to count the number of operands in a sequence or select them. Their use is not possible because a sequence becomes the arguments to these commands.

```
> s := x,y,z;
```

$$s := x, \ y, \ z$$

The command

```
> nops(s);
```

```
Error,
wrong number (or type) of parameters in function nops
```

is the same as the command

```
> nops(x,y,z);
```

```
Error,
wrong number (or type) of parameters in function nops
```

Here the arguments to the nops command are x, y, z, which constitute too many arguments. If you desire to count the number of operands in a sequence or select an operand from a sequence, you should first put the sequence in a list as follows

```
> nops([s]);
```

$$3$$

Alternatively, you can use the *selection operation* discussed in *Selection Operation* on page 146 to select the operands of a sequence.

Please note that many Maple commands return sequences. You may wish to put sequences into a list or set data structure. For example, when the arguments to the solve command are not sets, it returns a sequence of values if it finds multiple solutions.

```
> s := solve(x^4-2*x^3-x^2+4*x-2, x);
```

$$s := \sqrt{2}, \ -\sqrt{2}, \ 1, \ 1$$

The elements of the above sequence are values, not equations, because you did not use sets in the call to solve. Putting the solutions in a set removes duplicates.

```
> s := {s};
```

$$s := \{1, \ \sqrt{2}, \ -\sqrt{2}\}$$

The seq **Command**   The seq command creates sequences, a key tool for programming. *The* seq, add, *and* mul *Commands* on page 176 describes it in detail. The syntax takes either of the following general forms.

```
seq(f, i = a .. b)
seq(f, i = X)
```

Here *f*, *a*, *b*, and *X* are expressions and *i* is a name. In the first form, the expressions *a* and *b* must evaluate to two numerical constants or two single character strings. The result is the sequence produced by evaluating *f* after successively assigning the index *i* the values *a*, *a*+1, ..., *b*, (or up to the last value not exceeding *b*). If the value *a* is greater than *b* then the result is the NULL sequence.

```
> seq(i^2,i=1..4);
```

$$1, 4, 9, 16$$

```
> seq(i,i="d".."g");
```

$$\text{"d", "e", "f", "g"}$$

In the second form, seq(*f*, *i=X*), the result is the sequence produced by evaluating *f* after successively assigning the index *i* the operands of the expression *X* (or the individual characters, if *X* is a string). *Expression Trees: Internal Representation* on page 131 states the operands of a general expression.

```
> a := x^3+3*x^2+3*x+1;
```

$$a := x^3 + 3x^2 + 3x + 1$$

```
> seq(i,i=a);
```

$$x^3, 3x^2, 3x, 1$$

```
> seq(degree(i,x), i=a);
```

$$3, 2, 1, 0$$

```
> seq(i,i="maple");
```

$$\text{"m", "a", "p", "l", "e"}$$

The Dollar Operator   The sequence operator, $, also forms sequences. The primary purpose of $ is to represent a symbolic sequence such as x$n as in the following examples.

```
> diff(ln(x), x$n);
```

$$\frac{\partial}{\partial x^n} \ln(x)$$

```
> seq( diff(ln(x), x$n), n=1..5);
```

$$\frac{1}{x}, \ -\frac{1}{x^2}, \ \frac{2}{x^3}, \ -\frac{6}{x^4}, \ \frac{24}{x^5}$$

The general syntax of the $ operator is

```
f $ i = a .. b
f $ n
$ a .. b
```

where *f*, *a*, *b*, and *n* are expressions and *i* must evaluate to a name. In general, this operator is less efficient than seq and hence the seq command is preferred for programming.

In the first form, Maple creates a sequence by *substituting* the values *a*, *a*+1, ..., *b* for *i* in *f*.

The second form *f$n* is a shorthand notation for

```
f $ dummy = 1 .. n
```

where *dummy* is a dummy index variable. If the value of *n* is an integer, the result of the second form is the sequence consisting of the value of *f* repeated *n* times.

```
> x$3;
```

$$x, \ x, \ x$$

The third form *$a..b* is a shorthand notation for

```
dummy $ dummy = a .. b
```

If the values of *a* and *b* are numerical constants, this form is short for creating a numerical sequence *a*, *a*+1, *a*+2, ..., *b* (or up to the last value not exceeding *b*).

```
> $0..4;
```

$$0, \ 1, \ 2, \ 3, \ 4$$

The $ command differs from the seq command in that *a* and *b* do not need to evaluate to integers. However, when *a* and *b* do evaluate to specific values, seq is more efficient than $. See seq, add, *and* mul *Versus* $, sum, *and* product on page 177.

Sets and Lists

A *set* is an expression of the form

$$\boxed{\{ \; sequence \; \}}$$

and a *list* is an expression of the form

$$\boxed{[\; sequence \;]}$$

Note that a *sequence* may be empty, so {} represents the empty set and [] the empty list. A set is an *unordered* sequence of *unique* expressions. Maple removes duplicates and reorders the terms in a manner convenient for internal storage. A list is an *ordered* sequence of expressions with the order of the expressions specified by the user. Maple retains duplicate entries in a list.

```
> {y[1],x,x[1],y[1]};
```

$$\{x, \; y_1, \; x_1\}$$

```
> [y[1],x,x[1],y[1]];
```

$$[y_1, \; x, \; x_1, \; y_1]$$

A set is an expression of type set. Similarly, a list is an expression of type list. The operands in a list or set are the elements in the set or list. Select the elements of a list or set using either the op command or a subscript.

```
> t := [1, x, y, x-y];
```

$$t := [1, \; x, \; y, \; x - y]$$

```
> op(2,t);
```

$$x$$

```
> t[2];
```

$$x$$

Maple's ordering for sets is the order in which it stores the expressions in memory. The user should not make assumptions about this ordering. For example, in a different Maple session, the set above might appear in the ordering {y[1], x, x[1]}. You can sort elements of a list using the sort command.

Selection Operation The selection operation, [], selects components from an aggregate object. The aggregate objects include tables, arrays, sequences, lists, and sets. The syntax for the selection operation is

$$\boxed{name[\; sequence \;]}$$

If *name* evaluates to a table or array, Maple returns the table (array) entry.

```
> A := array([w,x,y,z]);
```

$$A := [w, \ x, \ y, \ z]$$

```
> A[2];
```

$$x$$

If *name* evaluates to a list, set, or sequence, and *sequence* evaluates to an integer, a range, or NULL, Maple performs a selection operation.

If *sequence* evaluates to an integer i, then Maple returns the ith operand of the set, list, or sequence. If *sequence* evaluates to a range, then Maple returns a set, list, or sequence containing the operands of the aggregate object as the range specifies. If *sequence* evaluates to NULL, then Maple returns a sequence containing all of the operands of the aggregate object.

```
> s := x,y,z:
> L := [s,s];
```

$$L := [x, \ y, \ z, \ x, \ y, \ z]$$

```
> S := {s,s};
```

$$S := \{y, \ z, \ x\}$$

```
> S[2];
```

$$z$$

```
> L[2..3];
```

$$[y, \ z]$$

```
> S[];
```

$$y, \ z, \ x$$

Negative integers count operands from the right.

```
> L := [t,u,v,w,x,y,z];
```

$$L := [t, \ u, \ v, \ w, \ x, \ y, \ z]$$

```
> L[-3];
```

$$x$$

```
> L[-3..-2];
```

$$[x, \ y]$$

You can also use select and remove to select elements from a list or set. See *The* map, select, *and* remove *Commands* on page 172.

Functions

A *function call* in Maple takes the form

$$\boxed{f(\ sequence\ )}$$

Often *f* will be a *name*, that is, the name of the function.

```
> sin(x);
```
$$\sin(x)$$

```
> min(2,3,1);
```
$$1$$

```
> g();
```
$$g()$$

```
> a[1](x);
```
$$a_1(x)$$

Maple executes a function call as follows. First, it evaluates *f* (typically yielding a procedure). Next, Maple evaluates the operands of *sequence* (the arguments) from left to right. (If any of the arguments evaluate to a sequence, Maple flattens the sequence of evaluated arguments into one sequence.) If *f* evaluated to a procedure, Maple invokes it on the argument sequence. Chapter 5 discusses this in detail.

```
> x := 1:
> f(x);
```
$$f(1)$$

```
> s := 2,3;
```
$$s := 2,\ 3$$

```
> f(s,x);
```
$$f(2,\ 3,\ 1)$$

```
> f := g;
```
$$f := g$$

```
> f(s,x);
```
$$g(2,\ 3,\ 1)$$

```
> g := (a,b,c) -> a+b+c;
```
$$g := (a,\ b,\ c) \rightarrow a + b + c$$

```
> f(s,x);
```
$$6$$

A function object's type is `function`. The operands are the arguments. The zeroth operand is the name of the function.

```
>  m := min(x,y,x,z);
```
$$m := \min(1, \ y, \ z)$$

```
> op(0,m);
```
$$min$$

```
> op(m);
```
$$1, \ y, \ z$$

```
> type(m,function);
```
$$true$$

```
> f := n!;
```
$$f := n!$$

```
> type(f, function);
```
$$true$$

```
> op(0, f);
```
$$factorial$$

```
> op(f);
```
$$n$$

In general, the function name *f* may be one of the following.

- name
- procedure definition
- integer
- float
- parenthesized algebraic expression
- function

Allowing *f* to be a procedure definition allows you to write, for example

```
> proc(t) t*(1-t) end(t^2);
```
$$t^2 \, (1 - t^2)$$

instead of

```
> h := proc(t) t*(1-t) end;
```

$$h := \mathbf{proc}(t)\, t \times (1 - t)\, \mathbf{end}$$

```
> h(t^2);
```

$$t^2 (1 - t^2)$$

If f is an integer or a float, Maple treats f as a constant operator. That is $f(x)$ returns f.

```
> 2(x);
```

$$2$$

The following rules define the meaning of a parenthesized algebraic expression.

```
> (f + g)(x), (f - g)(x), (-f)(x), (f@g)(x);
```

$$f(x) + g(x),\ f(x) - g(x),\ -f(x),\ f(g(x))$$

@ denotes functional composition; that is, f@g denotes $f \circ g$. These rules together with the previous rule mean that

```
> (f@g + f^2*g + 1)(x);
```

$$f(g(x)) + f(x)^2\, g(x) + 1$$

Notice that @@ denotes the corresponding exponentiation. That is, f@@n denotes $f^{(n)}$ which means f composed with itself n times.

```
> (f@@3)(x);
```

$$(f^{(3)})(x)$$

```
> expand(%);
```

$$f(f(f(x)))$$

Finally, f may be a function, as in

```
> cos(0);
```

$$1$$

```
> f(g)(0);
```

$$f(g)(0)$$

```
> D(cos)(0);
```

$$0$$

TABLE 4.11   The Arithmetic Operators

+	addition
–	subtraction
*	multiplication
&*	non-commutative multiplication
/	division
^	exponentiation
**	exponentiation

For more information on how to define a function, see chapter 5.

The Arithmetic Operators

Table 4.11 contains Maple's seven arithmetic operators. You may use all these items as binary operators. You may also use the operators + and – as prefix operators representing unary plus and unary minus. The two exponentiation operators ** and ^ are synonymous and you may use them interchangeably.

You can find the types and operands of the arithmetic operations listed below.

- The type of a sum or difference is +.
- The type of a product or quotient is * and the type of a power is ^.
- The operands of the sum $x - y$ are the terms x and $-y$.
- The operands of the product xy^2/z are factors x, y^2, and z^{-1}.
- The operands of the power x^a are the base x and the exponent a.

```
> whattype(x-y);
```
$$`+`$$
```
> whattype(x^y);
```
$$`^`$$

Arithmetic   Maple always computes the five arithmetic operations $x + y$, $x - y$, $x \times y$, x/y, and x^n, where n is an integer, if x and y are numbers. If the operands are floating-point numbers, Maple performs the arithmetic computation in floating-point.

```
> 2 + 3,  6/4,  1.2/7,  (2 + I)/(2 - 2*I);
```
$$5, \frac{3}{2}, .1714285714, \frac{1}{4} + \frac{3}{4} I$$
```
> 3^(1.2),  I^(1.0 - I);
```
$$3.737192819, 4.810477381\, I$$

The only other simplification done for numerical constants is reduction of fractional powers of integers and fractions. For integers n, m and fraction b,

$$(n/m)^b \to (n^b)/(m^b).$$

For integers n, q, r, d and fraction $b = q + r/d$ with $0 < r < d$,

$$n^b = n^{q+r/d} \to n^q \times n^{r/d}.$$

```
> 2^(3/2), (-2)^(7/3);
```

$$2\sqrt{2},\ 4(-2)^{1/3}$$

Automatic Simplifications   Maple automatically does these simplifications

```
> x - x,  x + x,  x + 0,  x*x,  x/x,  x*1,  x^0,  x^1;
```

$$0,\ 2x,\ x,\ x^2,\ 1,\ x,\ 1,\ x$$

for a symbol x, or an arbitrary expression. But these simplifications are not valid for all x. Some exceptions which Maple catches are

```
> infinity - infinity;
```

```
Error, invalid cancellation of infinity
```

```
> infinity/infinity;
```

```
Error, invalid cancellation of infinity
```

```
> 0/0;
```

```
Error, division by zero
```

```
> 0^0;
```

```
Error, 0^0 is undefined
```

In the following, n, m denote integers, a, b, c numerical constants, and x, y, z general symbolic expressions. Maple understands that addition and multiplication are associative and commutative, and so simplifies the following.

$$ax + bx \to (a + b)x$$

$$x^a \times x^b \to x^{a+b}$$

$$a(x + y) \to ax + ay$$

The first two simplifications mean that Maple adds like terms in poly-nomials automatically. The third means that Maple distributes numerical constants (integers, fractions, and floats) over sums, but does not do the same for non-numerical constants.

```
> 2*x + 3*x, x*y*x^2, 2*(x + y), z*(x + y);
```

$$5x, \ x^3 y, \ 2x + 2y, \ z(x+y)$$

The most difficult and controversial simplifications have to do with simplying powers x^y for non-integer exponents y.

Simplification of Repeated Exponentiation   In general, Maple does not do the simplification $(x^y)^z \rightarrow x^{(yz)}$ automatically because this procedure does not always provide an accurate answer. For example, letting $y = 2$ and $z = 1/2$, the first simplification would imply that $\sqrt{x^2} = x$, which is not necessarily true. Maple only does the first transformation above if it is provably correct for all complex x with the possible exception of a finite number of values, such as, 0 and ∞. Maple does $(x^a)^b \rightarrow x^{ab}$ if b is an integer, $-1 < a \leq 1$, or x is a positive real constant.

```
> (x^(3/5))^(1/2), (x^(5/3))^(1/2);
```

$$x^{3/10}, \ \sqrt{x^{5/3}}$$

```
> (2^(5/3))^(1/2), (x^(-1))^(1/2);
```

$$2^{5/6}, \ \sqrt{\frac{1}{x}}$$

Maple does not simplify $a^b c^b \rightarrow (ac)^b$ automatically, even if the answer is correct.

```
> 2^(1/2)+3^(1/2)+2^(1/2)*3^(1/2);
```

$$\sqrt{2} + \sqrt{3} + \sqrt{2}\sqrt{3}$$

The reason is that combining $\sqrt{2}\sqrt{3}$ to $\sqrt{6}$ would introduce a third unique square root. Calculating with roots is, in general, difficult and expensive, so Maple is careful not to create new roots. You may use the `combine` command to combine roots if you desire.

Non-Commutative Multiplication

The non-commutative multiplication operator `&*` acts as an inert operator (for example, the *neutral operators* described in *The Neutral Operators* on page 157), but the parser understands its binding strength to be equivalent to the binding strength of `*` and `/`.

The evalm command in the Maple Library interprets &* as the matrix multiplication operator. The evalm command also understands the form &*() as a generic matrix identity.

```
> with(linalg):
```

```
Warning, new definition for norm
Warning, new definition for trace
```

```
> A := matrix(2,2,[a,b,c,d]);
```

$$A := \left[\begin{array}{cc} a & b \\ c & d \end{array} \right]$$

```
> evalm( A &* &*() );
```

$$\left[\begin{array}{cc} a & b \\ c & d \end{array} \right]$$

```
> B := matrix(2,2,[e,f,g,h]);
```

$$B := \left[\begin{array}{cc} e & f \\ g & h \end{array} \right]$$

```
> evalm( A &* B - B &* A );
```

$$\left[\begin{array}{cc} bg - cf & af + bh - eb - fd \\ ce + dg - ga - hc & cf - bg \end{array} \right]$$

The Composition Operators

The composition operators are @ and @@. The @ operator represents function composition, that is, f@g in Maple denotes $f \circ g$.

```
> (f@g)(x);
```

$$f(g(x))$$

```
> (sin@cos)(Pi/2);
```

$$0$$

The @@ operator is the corresponding exponentiation operator representing repeated functional composition, that is, $f^{(n)}$ is denoted f@@n in Maple.

```
> (f@@2)(x);
```

$$(f^{(2)})(x)$$

```
> expand(%);
```

$$f(f(x))$$

```
> (D@@n)(f);
```

$$(D^{(n)})(f)$$

Unfortunately, mathematicians sometimes mix their notation in a context sensitive way. Usually $f^n(x)$ denotes composition; for example, D^n denotes the differential operator composed n times. Also $\sin^{-1}(x)$ denotes the inverse of the sine function, that is, composition to the power -1. But, sometimes mathematicians use $f^n(x)$ to denote ordinary powering, for example, $\sin^2(x)$ is the square of sine of x. Maple always uses $f^n(x)$ to denote repeated composition, and $f(x)^n$ to denote powering.

```
> sin(x)^2, (sin@@2)(x), sin(x)^(-1), (sin@@(-1))(x);
```

$$\sin(x)^2,\ (\sin^{(2)})(x),\ \frac{1}{\sin(x)},\ \arcsin(x)$$

The Ditto Operators

The value of the nullary operator, %, is the latest expression. The first and second expressions preceding the latest are the values of the nullary operators %% and %%%, respectively. The most common use of these operators is in an interactive Maple session where they refer to the previously computed results. The sequence of expressions defining these three nullary operators is the last three non-NULL values generated in the Maple session.

You may also use the ditto operators within the body of a Maple procedure. The ditto operators are local to the procedure. When you invoke the procedure, Maple initializes them to NULL and then updates them to refer to the last three non-NULL expressions that it computed during the execution of that particular procedure body.

The Factorial Operator

Maple uses the unary operator ! as a postfix operator which denotes the factorial function of its operand n. The input $n!$ is shorthand for the functional form `factorial(n)`.

```
> 0!, 5!, 2.5!;
```

$$1,\ 120,\ 2.5!$$

```
> (-2)!;
```

```
Error,
the argument to factorial should be non-negative
```

The type of an unevaluated factorial is !. Note that in Maple, $n!!$ does not denote the double factorial function. It denotes repeated factorial, $n!! = (n!)!$.

```
> 3!!;
```

$$720$$

The mod Operator

The mod operator evaluates an expression modulo m, for a non-zero integer m. That is, Maple writes $a \bmod m$ as a mod m. Maple uses one of two representations for an integer modulo m.

- In the *positive representation*, *integer* mod m is an integer between zero and $m-1$, inclusive. The following assignment selects the positive representation explicitly.

```
> 'mod' := modp;
```

This is the default representation.

- In the *symmetric representation*, *integer* mod m is an integer between -floor((abs(m)-1)/2) and floor(abs(m)/2). The following assignment selects the symmetric representation.

```
> 'mod' := mods;
```

Notice that you need back quotes around mod since it is a reserved word.

You may invoke the commands modp and mods directly if you desire. For example

```
> modp(9,5), mods(9,5);
```

$$4, -1$$

The mod operator understands the inert operator &^ for powering. That is, i&^j mod m calculates i^j mod m. Instead of first computing the integer i^j, which may be too large to compute, and then reducing modulo m, Maple computes the power using binary powering with remainder.

```
> 2^(2^100) mod 5;
```

Error, integer too large in context

```
> 2 &^ (2^100) mod 5;
```

$$1$$

The first operand of the mod operator may be a general expression. Maple evaluates the expression over the ring of integers modulo m. For polynomials, this means that it reduces rational coefficients modulo m. The mod operator knows many functions for polynomial and matrix arithmetic over finite rings and fields. For example, Factor for polynomial factorization, and Nullspace for matrix null-space.

```
> 1/2 mod 5;
```

$$3$$

```
> 9*x^2 + x/2 + 13 mod 5;
```

$$4 x^2 + 3 x + 3$$

```
> Factor(4*x^2 + 3*x + 3) mod 5;
```

$$4 (x + 3) (x + 4)$$

Do not confuse, for example, the commands factor and Factor. The former evaluates immediately; the latter is an inert command which Maple does not evaluate until you make the call to mod.

The mod command also knows how to compute over a Galois field $GF(p^k)$, that is, the finite field with p^k elements. See the ?mod on-line documentation for a list of the commands that mod knows, and for further examples.

The Neutral Operators

Maple possesses a *user-defined* or *neutral* operators facility. Form a neutral operator symbol using the ampersand character "&" followed by one or more characters. The two varieties of &-names depend on whether the sequence of characters is alphanumeric or non-alphanumeric:

- Any Maple *name* not requiring back quotes, preceded by the & character; for example, &wedge.

- The & character followed by one or more non-alphanumeric characters; for example, &+ or &++.

The following characters cannot appear in an &-name after the initial &:

& | ( ) [ ] { } ; : ' ' # \ %

as well as *newline* and *blank* characters.

Maple singles out the particular neutral operator symbol &* as a special token representing the non-commutative multiplication operator. The special property of &* is that the parser understands its binding strength to be equivalent to Maple's other multiplication operators. All other neutral operators have binding strength greater than the standard algebraic operators. See ?precedence for the order of precedence of all programming-language operators. See *Non-Commutative Multiplication* on page 153 which describes where to use &* in Maple.

You can use neutral operators as unary prefix operators, or infix binary operators, or as function calls. In any of these cases, they generate function calls with the name of the function being that of the neutral operator. (In the usual pretty-printing mode, these particular function calls are printed in binary operator format when exactly two operands exist and in unary operator format when exactly one operand exists, but the internal representation is an unevaluated function.) For example,

```
> a &~ b &~ c;
```

$$(a \mathbin{\&^\sim} b) \mathbin{\&^\sim} c$$

```
> op(%);
```

$$a \mathbin{\&^\sim} b, \ c$$

```
> op(0,%%);
```

$$\&^\sim$$

Maple imposes no semantics on the neutral operators. The user may define the operator to have a meaning by assigning the name to a Maple procedure. You can define manipulations on expressions containing such operators via Maple's interface to user-defined procedures for various standard library functions, including simplify, diff, combine, series, evalf, and many others. See *Extending Maple* on page 89.

Relations and Logical Operators

You can form new types of expression from ordinary algebraic expressions by using the *relational operators* <, >, <=, >=, =, and <>. The semantics of these operators is dependent on whether they occur in an *algebraic* context or in a *boolean* context.

In an algebraic context, the relational operators are simply place holders for forming equations or inequalities. Maple fully supports addition of equations or inequalities and multiplication of an equation or inequality by an algebraic expression. In the case of adding or subtracting two equations, for example, Maple applies the addition or subtraction to each side of the equations, thus yielding a new equation. In the case of multiplying an equation by an expression, Maple distributes the multiplication to each side of the equation. You may perform similar operations with inequalities.

```
> e  :=  x + 3*y = z;
```

$$e := x + 3\,y = z$$

```
> 2*e;
```

$$2\,x + 6\,y = 2\,z$$

The type of an equation is = or equation. An equation has two operands, the left-hand side, and the right-hand side. You can use the commands lhs and rhs to select the operands of an equation instead of op.

```
> op(0,e);
```

$$\text{`=`}$$

```
> lhs(e);
```

$$x + 3\,y$$

The type command also understands the types <>, <, and <=. Maple automatically converts inequalities involving > or >= to < and <=, respectively. All the relational types have two operands.

```
> e := a > b;
```

$$e := b < a$$

```
> op(e);
```

$$b,\ a$$

In a boolean context, Maple evaluates a relation to the value true or the value false. A boolean context includes the condition in an if statement and the condition in the while clause of a loop. You may also use the evalb command to evaluate a relation in a boolean context.

In the case of the operators <, <=, >, and >=, the difference of the operands must evaluate to a numeric constant and Maple compares this constant with zero.

```
> if 2<3 then "less" else "not less" fi;
```

$$\text{"less"}$$

In the case of the relations = and <>, the operands may be arbitrary expressions (algebraic or non-algebraic). This equality test for expressions deals only with syntactic equality of the Maple representations of the expressions, which is not the same as mathematical equivalence.

```
> evalb( x + y = y + x );
```

$$true$$

```
> evalb( x^2 - y^2  =  (x - y)*(x + y) );
```

$$false$$

In the latter example, applying the expand command results in an equation which evaluates to true.

```
> evalb( x^2 - y^2  =  expand( (x - y)*(x + y) ) );
```

$$true$$

You may use the is command instead of evalb to evaluate relations in a boolean context. The is command tries much harder than evalb to determine whether relations are true.

```
> is( x^2 - y^2  =  (x - y)*(x + y) );
```

$$true$$

```
> is( 3<Pi );
```

$$true$$

The Logical Operators Generally, you can form an expression using the *logical operators* and, or, and not, where the first two are binary operators and the third is a unary (prefix) operator. An expression containing one or more logical operators is automatically evaluated in a boolean context.

```
> 2>3 or not 5>1;
```

$$false$$

The precedence of the logical operators and, or, and not is analogous to multiplication, addition, and exponentiation, respectively. Here no parentheses are necessary.

```
> (a and b) or ((not c) and d);
```

$$a \text{ and } b \text{ or } \text{not } c \text{ and } d$$

The type names for the logical operators and, or, and not are and, or, and not, respectively. The first two have two operands, the latter one operand.

```
> b := x and y or z;
```

$$b := x \text{ and } y \text{ or } z$$

```
> whattype(b);
```

'or'

```
> op(b);
```

$$x \text{ and } y, z$$

Among operators of the same precedence, the evaluation of boolean expressions involving the logical operators and and or proceeds from left to right and terminates as soon as Maple can determine the truth of the whole expression. Consider the evaluation of the following.

```
a and b and c
```

If the result of evaluating *a* is false, you know that the result of the whole boolean expression will be false, regardless of what *b* and *c* evaluate to. These evaluation rules are commonly known as *McCarthy evaluation rules*. They are quite crucial for programming. Consider the following statement

```
if x <> 0 and f(x)/x > 1 then ... fi;
```

If Maple always evaluated both operands of the and clause, then when *x* is 0, evaluation would result in a division by zero error. The advantage of the above code is that Maple will only attempt to check the second condition when $x \neq 0$.

Boolean Expressions   In general, a boolean context requires a boolean expression. Use the boolean constants true, false, and FAIL, the *relational operators* and the *logical operators* for forming boolean expressions. The type command understands the name boolean to include all of these.

The evaluation of boolean expressions in Maple uses the following *three-valued logic*. In addition to the special names true and false, Maple also understands the special name FAIL. Maple sometimes uses the value FAIL as the value that a procedure returns when it is unable to completely solve a problem. In other words, you can view it as the value *don't know*.

```
> is(sin(1),positive);
```

true

```
> is(a-1,positive);
```

FAIL

In the context of the boolean clause in an if statement or a while statement, Maple determines the branching of the program by treating the

TABLE 4.12   Truth Tables

and	false	true	FAIL
false	false	false	false
true	false	true	FAIL
FAIL	false	FAIL	FAIL

or	false	true	FAIL
false	false	true	FAIL
true	true	true	true
FAIL	FAIL	true	FAIL

not	false	true	FAIL
	true	false	FAIL

value FAIL the same as the value false. Without three valued logic, whenever you use the is command you would have to test for FAIL separately. You would write

```
if is(a - 1, positive) = true then ...
```

The three valued logic allows you to write

```
if is(a - 1, positive) then ...
```

The evaluation of a boolean expression yields true, false, or FAIL according to table 4.12.

Note that three-valued logic leads to asymmetry in the use of if statements and while statements. For example, the following two statements are not equivalent.

```
if condition then statseq_1 else statseq_2 fi;
if not condition then statseq_2 else statseq_1 fi;
```

Depending on the desired action in the case where *condition* has the value FAIL, either the first or the second of these two if statements may be correct for a particular context.

Arrays and Tables

The table data type in Maple is a special data type for representing data in tables. Create a table either explicitly via the table command or implicitly by assignment to an indexed name. For example, the statements

```
> a := table([(Cu,1) = 64]);
```

$$a := \text{table}([$$
$$(Cu, 1) = 64$$
$$])$$

```
> a[Cu,1] := 64;
```

$$a_{Cu,1} := 64$$

have the same effect. They both create a `table` object with one component. The purpose of a table is to allow fast access to data with

```
> a[Cu,1];
```

$$64$$

The type of a table object is `table`. The first operand is the indexing function. The second operand is a list of the components. Note that tables and arrays have special evaluation rules; in order to access the table or array object, you must first apply the `eval` command.

```
> op(0,eval(a));
```

$$table$$

Table a has no indexing function and only one entry.

```
> op(1,eval(a));
> op(2,eval(a));
```

$$[(Cu, 1) = 64]$$

The `array` data type in Maple is a specialization of the `table` data type. An array is a `table` with specified dimensions, with each dimension an integer range. Create an array via the `array` command call.

```
> A := array(symmetric, 1..2, 1..2, [(1,1) = 3]);
```

$$A := \begin{bmatrix} 3 & A_{1,2} \\ A_{1,2} & A_{2,2} \end{bmatrix}$$

```
> A[1,2] := 4;
```

$$A_{1,2} := 4$$

```
> print(A);
```

$$\begin{bmatrix} 3 & 4 \\ 4 & A_{2,2} \end{bmatrix}$$

The ranges 1..2,1..2 specify two dimensions and bounds for the integers. You may include entries in the array command or insert them explicitly as shown. You may leave entries unassigned. In this example, the (2, 2) entry is unassigned.

```
> op(0,eval(A));
```

$$array$$

As for tables, the first operand is the indexing function (if any).

```
> op(1,eval(A));
```

$$symmetric$$

The second operand is the sequence of ranges.

```
> op(2,eval(A));
```

$$1..2, \ 1..2$$

The third operand is a list of entries.

```
> op(3, eval(A));
```

$$[(1, \ 1) = 3, \ (1, \ 2) = 4]$$

The example above displays only two entries in the array A since Maple knows the (2, 1) entry implicitly through the indexing function.

Series

The series data type in Maple represents an expression as a truncated power series with respect to a specified indeterminate, expanded about a particular point. Although you cannot input a series directly into Maple as an expression, you can create a series data type with the taylor or series commands which have the following syntax

```
taylor( f, x=a, n )
taylor( f, x )
series( f, x=a, n )
series( f, x )
```

If you do not specify the expansion point, it is by default $x = 0$. If you do not specify the order n, it is the value of the global variable Order, which by default is 6.

```
> s := series( exp(x), x=0, 4 );
```

$$s := 1 + x + \frac{1}{2}x^2 + \frac{1}{6}x^3 + O(x^4)$$

The type name for the series data type is `series`.

```
> type(s, series);
```

$$true$$

The zeroth operand is the expression $x - a$ where x denotes the specified indeterminate and a denotes the particular point of expansion.

```
> op(0, s);
```

$$x$$

The odd (first, third, …) operands are the coefficients of the series and the even operands are the corresponding integer exponents.

```
> op(s);
```

$$1,\ 0,\ 1,\ 1,\ \frac{1}{2},\ 2,\ \frac{1}{6},\ 3,\ O(1),\ 4$$

The coefficients may be general expressions but Maple restricts the exponents to *word-size* integers on the host computer, with a typical limit of nine or ten digits, ordered from least to greatest. Usually, the final pair of operands in the series data type are the special *order* symbol $O(1)$ and the integer n which indicates the order of truncation.

The `print` routine displays the final pair of operands using the notation $O(x^n)$ rather than more directly as $O(1)x^n$, where x is `op(0,s)`.

If Maple knows that the series is exact then it will not contain an order term. An example of this occurs when you apply the `series` command to a polynomial whose degree is less than the truncation degree for the series. A very special case is the *zero series*, which Maple immediately simplifies to the integer zero.

The `series` data structure represents generalized power series, which include Laurent series with finite principal parts. More generally, Maple allows the series coefficients to depend on x provided their growth is less than polynomial in x. $O(1)$ represents such a coefficient, rather than an arbitrary constant. An example of a non-standard generalized power series is

```
> series( x^x, x=0, 3 );
```

$$1 + \ln(x)\,x + \frac{1}{2}\ln(x)^2\,x^2 + O(x^3)$$

Maple can compute more general series expansions than the `series` data type supports. The Puisseux series is such an example. In these cases, the `series` command does not return a series data type, it returns a general algebraic expression.

```
> s := series( sqrt(sin(x)), x );
```

$$s := \sqrt{x} - \frac{1}{12}x^{5/2} + \frac{1}{1440}x^{9/2} + O(x^{11/2})$$

```
> type(s, series);
```

$$false$$

```
> type(s, '+');
```

$$true$$

Maple can also calculate with formal power series; see ?powseries.

Ranges

You often need to specify a *range* of numbers. For example, when you want to integrate a function over a range. In Maple, use the ellipsis operation to form ranges.

$$\boxed{expression\_1 \ .. \ expression\_2}$$

Specify the operator "`..`" using two *or more* consecutive periods. The ellipsis operator simply acts as a place holder in the same manner as using the relational operators in an algebraic context, primarily as a notational tool. A range has type "`..`" or range. A range has two operands, the left-limit and the right-limit, which you can access with the op, lhs, and rhs commands.

```
> r:=3..7;
```

$$r := 3..7$$

```
> op(0,r);
```

$$`..`$$

```
> lhs(r);
```

$$3$$

A typical application of ranges occurs in Maple's int, sum, and product commands. Interpret the operands of the ellipsis to mean the lower and upper limits of integration, summation, or products, respectively.

```
> int( f(x), x=a..b );
```

$$\int_a^b f(x)\,dx$$

You can use the range construct, with Maple's built-in command op, to extract a *sequence* of operands from an expression. The notation

$$op(a..b, c)$$

is equivalent to

$$seq(op(i,c),i=a..b)$$

For example,

```
> a := [ u, v, w, x, y, z ];
```

$$a := [u, v, w, x, y, z]$$

```
> op(2..5,a);
```

$$v, w, x, y$$

You can also use the range construct in combination with the concatenation operator to form a *sequence* as follows.

```
> x.(1..5);
```

$$x1, x2, x3, x4, x5$$

See *The Concatenation Operator* on page 118.

Unevaluated Expressions

Maple normally evaluates all expressions, but sometimes you need to tell Maple to delay evaluating an expression. An expression enclosed in a pair of single quotes

$$'expression'$$

is called an *unevaluated expression*. For example, the statements

```
> a := 1;  x := a + b;
```

$$a := 1$$
$$x := 1 + b$$

assign the value $1 + b$ to the name x, while the statements

```
> a := 1;  x := 'a' + b;
```

$$a := 1$$
$$x := a + b$$

assign the value $a + b$ to the name x if b has no value.

The effect of evaluating a quoted expression is to strip off one level of quotes, so in some cases nested levels of quotes are very useful. Note the distinction between *evaluation* and *simplification* in the statement

```
> x := '2 + 3';
```

$$x := 5$$

which assigns the value 5 to the name x even though this expression contains quotes. The evaluator simply strips off the quotes, but the *simplifier* transforms the expression $2 + 3$ into the constant 5.

The result of evaluating an expression with two levels of quotes is an expression of type uneval. This expression has only one operand, namely the expression inside the outermost pair of quotes.

```
> op(''x - 2'');
```

$$x - 2$$

```
> whattype(''x - 2'');
```

$$uneval$$

A special case of unevaluation arises when a name, which Maple may have assigned a value, needs unassigning so that in the future the name simply stands for itself. You can accomplish this by assigning the quoted name to itself.

```
> x := 'x';
```

$$x := x$$

Now x stands for itself as if Maple had never assigned it a value.

Another special case of unevaluation arises in the function call

$$\boxed{\text{'f' } (sequence)}$$

Suppose the arguments evaluate to the sequence a. Since the result of evaluating $'f'$ is not a procedure, Maple returns the unevaluated function call $f(a)$.

```
> ''sin''(Pi);
```

$$\text{'sin'}(\pi)$$

```
> %;
```

$$\sin(\pi)$$

```
> %;
```

$$0$$

You will find this facility useful when writing procedures which implement simplification rules. See *Extending Certain Commands* on page 96.

Constants

Maple has a general concept of *symbolic constants*, and initially assigns the global variable `constants` the following expression sequence of names

```
> constants;
```

$$\textit{false}, \gamma, \infty, \textit{true}, \textit{Catalan}, \textit{FAIL}, \pi$$

implying that Maple understands these particular names to be of type `constant`. The user may define additional names (specifically, they must be the simplest type of names called *strings*—see *Names* on page 117) to be constants by redefining the value of this global variable.

```
> type(g,constant);
```

$$\textit{false}$$

```
> constants := constants, g;
```

$$\textit{constants} := \textit{false}, \gamma, \infty, \textit{true}, \textit{Catalan}, \textit{FAIL}, \pi, g$$

```
> type(g,constant);
```

$$\textit{true}$$

Generally, a Maple expression is of type `constant` if it is of type `numeric`, or one of the constants, or an unevaluated function with all arguments of type `constant`, or a sum, product, or power with all operands of type `constant`. For example, the following expressions are of type `constant`: `2, sin(1), f(2,3), exp(gamma), 4+Pi, 3+I, 2*gamma/Pi^(1/2)`

Structured Types

Sometimes a simple type check does not give enough information. For example, the command

```
> type( x^2, '^' );
```

$$\textit{true}$$

tells you that `x^2` is an exponentiation but it does not tell you whether or not the exponent is, say, an integer. In such cases, you need *structured types*.

```
> type( x^2, name^integer );
```

$$true$$

Since x is a name and 2 is an integer, the command returns true. The square root of x does not have this type.

```
> type( x^(1/2), name^integer );
```

$$false$$

The expression (x+1)^2 does not have type name^integer, because x+1 is not a name.

```
> type( (x+1)^2, name^integer );
```

$$false$$

The type anything matches any expression.

```
> type( (x+1)^2, anything^integer );
```

$$true$$

An expression matches a set of types if the expression matches one of the types in the set.

```
> type( 1, {integer, name} );
```

$$true$$

```
> type( x, {integer, name} );
```

$$true$$

The type set(*type*) matches a set of elements of type *type*.

```
> type( {1,2,3,4}, set(integer) );
```

$$true$$

```
> type( {x,2,3,y}, set( {integer, name} ) );
```

$$true$$

Similarly, the type list(*type*) matches a list of elements of type *type*.

```
> type( [ 2..3, 5..7 ], list(range) );
```

$$true$$

Note that e^2 is not of type anything^2.

```
> exp(2);
```

$$e^2$$

```
> type( %, anything^2 );
```

$$false$$

The reason is that e^2 is simply the pretty-printed version of exp(2).

```
> type( exp(2), 'exp'(integer) );
```

$$true$$

You should use single quotes (') around Maple commands in type expressions to delay evaluation.

```
> type( int(f(x), x), int(anything, anything) );
```

```
Error, testing against an invalid type
```

Here Maple evaluated int(anything, anything) and got

```
> int(anything, anything);
```

$$\frac{1}{2}\,anything^2$$

which is not a valid type. If you put single quotes around the int command, the type checking works as intended.

```
> type( int(f(x), x), 'int'(anything, anything) );
```

$$true$$

The type specfunc(*type*, *f*) matches the function *f* with zero or more arguments of type *type*.

```
> type( exp(x), specfunc(name, exp) );
```

$$true$$

```
> type( f(), specfunc(name, f) );
```

$$true$$

The type function(*type*) matches any function with zero or more arguments of type *type*.

```
> type( f(1,2,3), function(integer) );
```

$$true$$

```
> type( f(1,x,Pi), function( {integer, name} ) );
```

$$true$$

You can also test the number (and types) of arguments. The type anyfunc(*t1*, ..., *tn*) matches any function with *n* arguments of the listed types.

```
> type( f(1,x), anyfunc(integer, name) );
```

$$true$$

```
> type( f(x,1), anyfunc(integer, name) );
```

$$false$$

Another useful variation is to use the And, Or, and Not type constructors to create boolean combinations of types.

```
> type(Pi, 'And( constant, numeric)');
```

$$false$$

```
> type(Pi, 'And( constant, Not(numeric))');
```

$$true$$

See ?type,structured for more information on structured types or ?type,definition on how to define your own types.

4.5 Useful Looping Constructs

The Repetition Statement on page 125 describes the for loop and while loop. Many common kinds of loops appear so often that Maple provides special purpose commands for them. These commands help to make writing programs simpler and more efficient. They are the "bread and butter" commands in the Maple language. You can group the seven loop-like commands in Maple into three categories

1. map, select, remove
2. zip
3. seq, add, mul

The map, select, and remove Commands

The map command applies a function to every element of an aggregate object. The simplest form of the map command is

$$\boxed{\text{map}(f, x)}$$

where *f* is a function and *x* is an expression. The map command replaces each operand x_i of the expression x with $f(x\_i)$.[1]

---
[1] Exception: for a table or array, Maple applies the function to the entries of the table or array, and not to the operands or indices.

```
> map( f, [a,b,c] );
```

$$[f(a),\ f(b),\ f(c)]$$

For example, if you have a list of integers, create a list of their absolute values and of their squares using the map command.

```
> L := [ -1, 2, -3, -4, 5 ];
```

$$L := [-1,\ 2,\ -3,\ -4,\ 5]$$

```
> map(abs,L);
```

$$[1,\ 2,\ 3,\ 4,\ 5]$$

```
> map(x->x^2,L);
```

$$[1,\ 4,\ 9,\ 16,\ 25]$$

The general syntax of the map command is

```
map( f, x, y1, ..., yn )
```

where f is a function, x is any expression, and $y1, \ldots, yn$ are expressions. The action of map is to replace each operand x_i of x by $f(x\_i,\ y1,\ \ldots,\ yn)$.

```
> map( f, [a,b,c], x, y );
```

$$[f(a,\ x,\ y),\ f(b,\ x,\ y),\ f(c,\ x,\ y)]$$

```
> L := [ seq(x^i, i=0..5) ];
```

$$L := [1,\ x,\ x^2,\ x^3,\ x^4,\ x^5]$$

```
> map( (x,y)->x^2+y, L, 1);
```

$$[2,\ x^2 + 1,\ x^4 + 1,\ x^6 + 1,\ x^8 + 1,\ x^{10} + 1]$$

Note that the result type may not necessarily be the same as the input type for algebraic types because of simplification. Consider the following examples.

```
> a := 2-3*x+I;
```

$$a := 2 - 3x + I$$

```
> map( z->z+x, a);
```

$$2 + I$$

Adding an x to each term of a yields 2+x-3*x+x+I-x = 2+I, which is a complex number.

```
> type(%, complex);
```

$$true$$

However, a is not a complex number.

```
> type(a, complex);
```

$$false$$

The `select` and `remove` commands have the same syntax as the `map` command and they work in a similar way. The simplest forms are

```
select( f, x )
remove( f, x )
```

where *f* is a boolean-valued function and *x* is an expression which must be one of a sum, product, list, set, function, or indexed name.

The `select` command selects the operands of *x* which satisfy the boolean-valued function *f*, creating a new object of the same type as *x*. Maple discards those operands for which *f* does not return `true`.

The `remove` command does the opposite of `select`. It removes the operands of *x* that satisfy *f*.

```
> X := [seq(i,i=1..10)];
```

$$X := [1, 2, 3, 4, 5, 6, 7, 8, 9, 10]$$

```
> select(isprime,X);
```

$$[2, 3, 5, 7]$$

```
> remove(isprime,X);
```

$$[1, 4, 6, 8, 9, 10]$$

The general forms of the `select` and `remove` commands are

```
select( f, x, y1, ..., yn )
remove( f, x, y1, ..., yn )
```

where *f* is a function, *x* is a sum, product, list, set, function or indexed name, and *y1*, ..., *yn* are expressions. As with the general form of the `map` command the expressions *y1*, ..., *yn* are passed to the function *f*.

```
> X := {2, sin(1), exp(2*x), x^(1/2)};
```

$$X := \{2, \sin(1), \sqrt{x}, e^{(2x)}\}$$

```
> select(type, X, function);
```

$$\{\sin(1), e^{(2x)}\}$$

```
> remove(type, X, constant);
```

$$\{\sqrt{x},\ e^{(2x)}\}$$

```
> X := 2*x*y^2 - 3*y^4*z + 3*z*w + 2*y^3 - z^2*w*y;
```

$$X := 2\,x\,y^2 - 3\,y^4\,z + 3\,z\,w + 2\,y^3 - z^2\,w\,y$$

```
> select(has, X, z);
```

$$-3\,y^4\,z + 3\,z\,w - z^2\,w\,y$$

```
> remove( x -> degree(x)>3, X );
```

$$2\,x\,y^2 + 3\,z\,w + 2\,y^3$$

The zip Command

Use the zip command to merge two lists or vectors. The zip command has two forms

```
zip(f, u, v)
zip(f, u, v, d)
```

where f is a binary function, u and v are both lists or vectors, and d is a value. The action of zip is for each pair of operands u_i, v_i, to create a new list or vector out of $f(u\_i, v\_i)$. The following is an example of the action of zip.

```
> zip( (x,y)->x.y, [a,b,c,d,e,f], [1,2,3,4,5,6] );
```

$$[a1,\ b2,\ c3,\ d4,\ e5,\ f6]$$

If the lists or vectors are not the same length, the length of the result depends on whether you provide d. If you do not specify d, the length will be the length of the smaller of u and v.

```
> zip( (x,y)->x+y, [a,b,c,d,e,f], [1,2,3] );
```

$$[a + 1,\ b + 2,\ c + 3]$$

If you do specify d, the length of the result of the zip command will be the length of the longer list (or vector) and Maple uses d for the missing value(s).

```
> zip( (x,y)->x+y, [a,b,c,d,e,f], [1,2,3], xi );
```

$$[a + 1,\ b + 2,\ c + 3,\ d + \xi,\ e + \xi,\ f + \xi]$$

Note that Maple does *not* pass the extra argument, xi, to the function f as it does with the map command.

The seq, add, and mul Commands

The seq, add, and mul commands form sequences, sums, and products, respectively. Use the following syntax with these commands.

```
seq(f, i = a..b)
add(f, i = a..b)
mul(f, i = a..b)
```

where f, a, and b are expressions and i is a name. The expressions a and b must evaluate to numerical constants (except in the case of seq, where they may be single character strings).

The result of seq is the sequence that Maple produces by evaluating f after successively assigning the index name i the values a, $a+1$, ..., b, (or up to the last value not exceeding b). The result of add is the sum of the same sequence, and the result of mul is the product of the sequence. If the value a is greater than b, the result is the NULL sequence, 0, and 1, respectively.

```
> seq(i^2,i=1..4);
```

$$1, 4, 9, 16$$

```
> mul(i^2,i=1..4);
```

$$576$$

```
> add(x[i], i=1..4);
```

$$x_1 + x_2 + x_3 + x_4$$

```
> mul(i^2, i = 4..1);
```

$$1$$

```
> seq(i, i = 4.123 .. 6.1);
```

$$4.123, 5.123$$

You may also use the commands seq, add, and mul with the following syntax.

```
seq(f, i = X)
add(f, i = X)
mul(f, i = X)
```

where f and X are expressions and i is a name.

The result of seq in this form is the sequence that Maple produces by evaluating f after successively assigning the operands of the expression X (or the characters of string X) to the index i. The result of add is the sum of the same sequence, and the result of `mul` is the product of the same sequence.

```
> a := x^3 + 3*x^2 + 3*x + 1;
```

$$a := x^3 + 3x^2 + 3x + 1$$

```
> seq(degree(i,x), i=a);
```

$$3, 2, 1, 0$$

```
> seq(i, i="square");
```

$$\text{``s'', ``q'', ``u'', ``a'', ``r'', ``e''}$$

```
> add(degree(i,x), i=a);
```

$$6$$

```
> a := [23,-42,11,-3];
```

$$a := [23, -42, 11, -3]$$

```
> mul(abs(i),i=a);
```

$$31878$$

```
> add(i^2,i=a);
```

$$2423$$

seq, add, and mul Versus $, sum, and product   Note that the dollar operator, $, and the `sum` and `product` commands are very similar to the `seq`, `mul`, and `add` commands. However, they differ in an important way. The index variable i and the end points a and b do not need to be integers. For example

```
> x[k] $ k=1..n;
```

$$x_k \ \$ \ (k = 1..n)$$

The design of these commands is for *symbolic* sequences, sums, and products. As with the `int` (integration) command, the index variable k is a global variable to which you must not assign a value.

When should you use `seq`, `add`, `mul` verses $, `sum`, and `product`?

When you are computing a symbolic sum or product. For example, if the end points are unknowns, then clearly you must use $, `sum`, and `product`. When you are computing an explicit finite sequence, sum, or product, that is, you know that the range points a and b are integers, then use `seq`, `add`, or `mul`. These latter commands are more efficient than their symbolic counterparts $, `sum`, and `product`.

4.6 Substitution

The subs command does a *syntactic* substitution. It replaces subexpressions in an expression with a new value; the subexpressions must be operands in the sense of the op command.

```
> expr := x^3 + 3*x + 1;
```

$$expr := x^3 + 3\,x + 1$$

```
> subs(x=y, expr);
```

$$y^3 + 3\,y + 1$$

```
> subs(x=2, expr);
```

$$15$$

The syntax of the subs command is

$$\boxed{\text{subs(}s,\ expr\ )}$$

where *s* is either an equation, a list, or set of equations. Maple traverses the expression *expr* and compares each operand in *expr* with the left-hand side(s) of the equation(s) *s*. If an operand is equal to a left-hand side of an equation in *s*, then subs replaces the operand with the right-hand side of the equation. If *s* is a list or set of equations, then Maple makes the substitutions indicated by the equations simultaneously.

```
> f := x*y^2;
```

$$f := x\,y^2$$

```
> subs( {y=z, x=y, z=w}, f );
```

$$y\,z^2$$

The general syntax of the subs command is

$$\boxed{\text{subs(}s1,\ s2,\ \ldots,\ sn,\ expr\ )}$$

where *s1*, *s2*, ..., *sn* are equations or sets or lists of equations, $n > 0$, and *expr* is an expression. This is equivalent to the following sequence of substitutions.

$$\boxed{\text{subs(}sn,\ \ldots,\ \text{subs(}s2,\ \text{subs(}s1,\ expr\ )\ )\ )}$$

Thus, subs substitutes according to the given equations from left to right. Notice the difference between the previous example and the following one.

```
> subs( y=z, x=y, z=w, f );
```

$$y\,w^2$$

Maple does not evaluate the result of a substitution.

```
> subs( x=0, sin(x) + x^2 );
```

$$\sin(0)$$

If you want to combine the acts of substitution and evaluation, use the two-parameter version of the `eval` command instead of subs.

```
> eval(sin(x) + x^2, x=0);
```

$$0$$

Substitution only compares operands in the expression tree of *expr* with the left-hand side of an equation.

```
> subs(a*b=d, a*b*c);
```

$$a\,b\,c$$

The substitution did not result in d*c as intended because the operands of the product a*b*c are a, b, c. That is, the products a*b, b*c, and a*c do not appear explicitly as operands in the expression a*b*c; consequently, subs does not see them.

The easiest way to make such substitutions is to solve the equation for one unknown and substitute for that unknown, that is

```
> subs(a=d/b, a*b*c);
```

$$d\,c$$

You cannot always do this, and you may find that it does not always produce the results you expect. The `algsubs` routine provides a more powerful substitution facility.

```
> algsubs(a*b=d, a*b*c);
```

$$d\,c$$

Note also that operands of a rational power $x^{n/d}$ are x and n/d. Although in the following example

```
> subs( x^(1/2)=y, a/x^(1/2) );
```

$$\frac{a}{\sqrt{x}}$$

it looks as though the output has a $\sqrt{x}$ in it, the operands of this expression are a and $x^{-1/2}$. Think of the division as a negative power in a product, that is, $a \times x^{-1/2}$. Because the operands of $x^{-1/2}$ are x and $-1/2$, subs does not see $x^{1/2}$ in $x^{-1/2}$. The solution is to substitute for the negative power $x^{-1/2}$.

```
> subs( x^(-1/2)=1/y, a/x^(1/2) );
```

$$\frac{a}{y}$$

The reader should refer to the on-line help information under ?algsubs for more details. Note that the algsubs command, as powerful as it is, is also much more computationally expensive than the subs command.

4.7   Conclusion

This chapter discusses the elements of Maple's language. Maple breaks your input into its smallest meaningful parts, called tokens. Its language statements include assignments, conditional, looping, and reading from and saving to files. Many types of expressions exist within Maple, and using its expression trees tells you of the type and operands in an expression. You have seen the efficient looping constructs map, zip, and seq, and how to make substitutions.

CHAPTER

5

Procedures

The proc command defines procedures in Maple. This chapter describes the syntax and semantics of the proc command in the same manner as chapter 4 describes the rest of the Maple programming language. This chapter explains the concepts of local and global variables and how Maple passes arguments to procedures. The chapter also provides exercises to help extend your understanding of Maple procedures.

5.1 Procedure Definitions

A Maple procedure definition has the following general syntax.

```
proc( P )
local L;
global G;
options O;
description D;
B
end
```

Here, *B* is a sequence of statements forming the body of the procedure. The formal parameters, *P*, along with the local, global, options, and description clauses are all optional.

The following is a simple Maple procedure definition. It has two *formal parameters*, *x* and *y*, no local, global, options, or description clauses, and only one statement in the body of the procedure.

```
> proc(x,y)
>     x^2 + y^2
> end;
```

$$\textbf{proc}(x, \ y) \, x^2 + y^2 \ \textbf{end}$$

You can give a name to a procedure as to any other Maple object.

```
> F := proc(x,y) x^2 + y^2 end;
```

$$F := \textbf{proc}(x, \ y) \, x^2 + y^2 \ \textbf{end}$$

You can then *execute* (invoke) it using the function call,

$$\boxed{F \ (\ A \ )}$$

When Maple executes the statements of the body of a procedure, it replaces the formal parameters, P, with the actual parameters, A, from the function call. Note that Maple evaluates the actual parameters, A, before substituting them for the formal parameters, P.

Normally, the result a procedure returns after executing is the value of the last executed statement from the body of the procedure.

```
> F(2,3);
```

$$13$$

Mapping Notation

You can also define simple one-line procedures using an alternate syntax borrowed from algebra.

$$\boxed{(\ P \ ) \ \text{->} \ B}$$

The sequence, P, of formal parameters may be empty and the body, B, of the procedure must be a single expression or an `if` statement.

```
> F := (x,y) -> x^2 + y^2;
```

$$F := (x, \ y) \rightarrow x^2 + y^2$$

If your procedure involves only one parameter, then you may omit the parentheses around the formal parameter.

```
> G := n -> if n<0 then 0 else 1 fi;
```

$$G := \textbf{proc}(n)$$

$$\textbf{option } \textit{operator, arrow;}$$

$$\textbf{if } n < 0 \textbf{ then } 0 \textbf{ else } 1 \textbf{ fi}$$

end

```
> G(9), G(-2);
```

$$1, 0$$

The intended use for the *arrow notation* is solely for simple one-line function definitions. It does not provide a mechanism for specifying local or global variables, or options.

Unnamed Procedures and Their Combinations

Procedure definitions are valid Maple expressions. You can create, manipulate, and invoke all of them without assigning to a name.

```
> (x) -> x^2;
```

$$x \to x^2$$

You invoke an unnamed procedure in the following manner.

```
> ( x -> x^2 )( t );
```

$$t^2$$

```
> proc(x,y) x^2 + y^2 end(u,v);
```

$$u^2 + v^2$$

A common use of unnamed procedures occurs in conjunction with the map command.

```
> map( x -> x^2, [1,2,3,4] );
```

$$[1, 4, 9, 16]$$

You can add procedures together, or, if appropriate, you can process them using commands, such as the differential operator, D.

```
> D(x -> x^2);
```

$$x \to 2x$$

```
> F := D(exp + 2*ln);
```

$$F := \exp + 2 \left(a \to \frac{1}{a} \right)$$

You can apply the result, F, directly to arguments.

Procedure Simplification

When you create a procedure, Maple does not evaluate the procedure but it does *simplify* the body of the procedure.

```
> proc(x) local t;
>     t := x*x*x + 0*2;
>     if true then sqrt(t); else t^2 fi;
> end;
```

$$\mathbf{proc}(x)\ \mathbf{local}\ t;\ t := x^3;\ \text{sqrt}(t)\ \mathbf{end}$$

Maple simplifies procedures with the operator option even further.

```
> x -> 3/4;
```

$$\frac{3}{4}$$

```
> (x,y,z) -> h(x,y,z);
```

$$h$$

Procedure simplification is a simple form of program optimization.

5.2   Parameter Passing

Consider what happens when Maple evaluates a function or procedure.

$$\boxed{F(\ \textit{ArgumentSequence}\ )}$$

First, Maple evaluates *F*. Then it evaluates the *ArgumentSequence*. If any of the arguments evaluate to a sequence, Maple flattens the resulting sequence of sequences out into a single sequence, the sequence of *actual parameters*. Suppose *F* evaluates to a procedure.

$$\boxed{\begin{array}{l} \text{proc}(\ \textit{FormalParameters}\ ) \\ \textit{body}; \\ \text{end}; \end{array}}$$

Maple then executes the statements in the *body* of the procedure, substituting the actual parameters for the formal parameters.

Consider the following example.

```
> s := a,b: t := c:
> F := proc(x,y,z) x + y + z end:
> F(s,t);
```

$$a + b + c$$

Here, s,t is the *argument sequence*, a,b,c is the *actual parameter sequence* and x,y,z is the *formal parameter sequence*.

The number of actual parameters, *n*, may differ from the number of formal parameters. If too few actual parameters exist, then an error occurs if (and only if) a missing parameter is actually used during the execution of the procedure body. Maple simply ignores extra parameters.

```
> f := proc(x,y,z) if x>y then x else z fi end:
> f(1,2,3,4);
```

$$3$$

```
> f(1,2);

Error, (in f) f uses a 3rd argument, z,
which is missing

> f(2,1);
```

$$2$$

Declared Parameters

You may write procedures that only work for certain types of input. Use *declared formal parameters* so that when you use the procedure with the wrong types of input Maple issues an informative error message. A type declaration has the following syntax.

> *parameter* :: *type*

Here *parameter* is the name of the formal parameter, and *type* is a type. Maple knows many types of expressions; see ?type.

When invoking the procedure, before executing the body of the procedure, Maple tests the types of the actual parameters from left to right. Any of these tests may generate an error message. If no type error occurs, the procedure executes.

```
> MAX := proc(x::numeric, y::numeric)
>    if x>y then x else y fi
> end:
> MAX(Pi,3);

Error, MAX expects its 1st argument, x, to be of type
numeric, but received Pi
```

You can also use declared parameters with the operator option.

```
> G := (n::even) -> n! * (n/2)!;
```

$$G := n::even \rightarrow n! \left(\frac{1}{2}n\right)!$$

```
> G(6);
```

$$4320$$

```
> G(5);
```

```
Error, G expects its 1st argument, n, to be of type
even, but received 5
```

If you do not declare the type of a parameter, it can have any type. Thus, proc(x) is equivalent to proc(x::anything). If that is what you intend, you should use the latter form to inform other users that you intend your procedure to work for any input.

The Sequence of Arguments

You do not need to supply names for the formal parameters. You can access the entire sequence of actual arguments from within the procedure, using the name args. The following procedure simply builds a list of its arguments.

```
> f := proc() [args] end;
```

$$f := \mathbf{proc}() \,[args] \,\mathbf{end}$$

```
> f(a,b,c);
```

$$[a, b, c]$$

```
> f(c);
```

$$[c]$$

```
> f();
```

$$[]$$

The ith argument is simply args[i]. Thus, the following two procedures are equivalent, provided you call them with at least two actual parameters of type numeric.

```
> MAX := proc(x::numeric,y::numeric)
>    if x > y then x else y fi;
> end;
```

$$MAX := \mathbf{proc}(x::numeric \,,\, y::numeric)$$

$$\textbf{if } y < x \textbf{ then } x \textbf{ else } y \textbf{ fi}$$

$$\textbf{end}$$

```
> MAX := proc()
>    if args[1] > args[2] then args[1] else args[2] fi;
> end;
```

$$MAX := \textbf{proc}()$$

$$\textbf{if } \text{args}_2 < \text{args}_1 \textbf{ then } \text{args}_1 \textbf{ else } \text{args}_2 \textbf{ fi}$$

$$\textbf{end}$$

The nargs command provides the total number of actual parameters. This allows you to easily write a procedure, MAX, which finds the maximum of any number of arguments.

```
> MAX := proc()
>    local i,m;
>    if nargs = 0 then RETURN(FAIL) fi;
>    m := args[1];
>    for i from 2 to nargs do
>        if args[i] > m then m := args[i] fi;
>    od;
>    m;
> end:
```

The maximum of the three values 2/3, 1/2, and 4/7 is

```
> MAX(2/3, 1/2, 4/7);
```

$$\frac{2}{3}$$

5.3 Local and Global Variables

Variables inside a procedure are either local to that procedure or global. Variables outside procedures are global. Maple considers local variables in different procedure invocations to be different variables, even if they have the same name. Thus, a procedure can change the value of a local variable without affecting variables of the same name in other procedures or a global variable of the same name. You should always declare which variables are local and which are global in the following manner.

```
local L1, L2, ..., Ln;
global G1, G2, ..., Gm;
```

In the procedure below, *i* and *m* are local variables.

```
> MAX := proc()
>    local i,m;
>    if nargs = 0 then RETURN(0) fi;
>    m := args[1];
>    for i from 2 to nargs do
>        if args[i] > m then m := args[i] fi;
>    od;
>    m;
> end:
```

In the case of nested procedures, where one procedure is defined within the body of another, variables can also acquire local or global declaration from procedures which enclose them. See *Nested Procedures* on page 49 for details and examples of nested procedures.

If no declaration is made of whether a variable is local or global, Maple decides. A variable is automatically made local if:

• It appears on the left-hand side of an assignment statement, for example A in A := y or A[1] := y.

• It appears as the index variable in a for loop, or in a seq, add, or mul command.

If neither of these two rules applies, the variable is a global variable.

```
> MAX := proc()
>    if nargs = 0 then RETURN(0) fi;
>    m := args[1];
>    for i from 2 to nargs do
>        if args[i] > m then m := args[i] fi;
>    od;
>    m;
> end:
```

```
Warning, 'm' is implicitly declared local
Warning, 'i' is implicitly declared local
```

Maple declares *m* local because it appears on the left-hand side of the assignment m:=args[1], and *i* local because it is the index variable of a for loop.

Do not rely on this facility to declare local variables. Declare all your local variables explicitly. Rely instead on the warning messages to help you identify variables that you have misspelled or have forgotten to declare.

The newname procedure below creates the next unused name in the sequence $C1, C2, \ldots$. The name that newname creates is a global variable since neither of the two rules above apply to C.N.

```
> newname := proc()
>    global N;
>    N := N+1;
>    while assigned(C.N) do
>         N := N+1;
>    od;
>    C.N;
> end:
> N := 0;
```

$$N := 0$$

The `newname` procedure does not take any arguments.

```
> newname() * sin(x) + newname() * cos(x);
```

$$C1\sin(x) + C2\cos(x)$$

Assigning values to global variables inside procedures is generally a poor idea. Any change of the value of a global variable affects all uses of the variable, even any of which you were unaware. Thus, you should only use this technique judiciously.

Evaluation of Local Variables

Local variables are special in another very important way. During the execution of a procedure body, they evaluate exactly *one level*. Maple evaluates global variables fully, even inside a procedure.

This section should help to clarify this concept. Consider the following examples.

```
> f := x + y;
```

$$f := x + y$$

```
> x := z^2/ y;
```

$$x := \frac{z^2}{y}$$

```
> z := y^3 + 3;
```

$$z := y^3 + 3$$

The normal full recursive evaluation yields

```
> f;
```

$$\frac{(y^3 + 3)^2}{y} + y$$

You can control the actual level of evaluation by using `eval`. Using the following sequence of commands, you can evaluate to one level, two levels, and three levels.

```
> eval(f,1);
```

$$x + y$$

```
> eval(f,2);
```

$$\frac{z^2}{y} + y$$

```
> eval(f,3);
```

$$\frac{(y^3 + 3)^2}{y} + y$$

The notion of the use of *one-level evaluation*[1] is important for efficiency. It has very little effect on the behavior of programs because when you are programming, you tend to write code in an organized sequential fashion. In the rare case where a procedure body requires a full-recursive evaluation of a local variable, you may use the `eval` command.

```
> F := proc()
>    local x, y, z;
>    x := y^2;   y := z;   z := 3;
>    eval(x)
> end:
> F();
```

$$9$$

Without the call to `eval`, the answer would be y^2.

You can still use local variables as unknowns just like global variables. For example, in the following procedure, the local variable x does not have an assigned value. The procedure uses it as the variable in the polynomial $x^n - 1$.

```
> RootsOfUnity := proc(n)
>    local x;
>    [solve( x^n - 1=0, x )];
> end:
```

---

[1] Such a concept of evaluation does not occur in traditional programming languages. However, here, you may assign to a variable a formula involving other variables, which in turn you may assign values and so on.

```
> RootsOfUnity(5);
```

$$[1, \frac{1}{4}\sqrt{5} - \frac{1}{4} + \frac{1}{4}I\sqrt{2}\sqrt{5+\sqrt{5}},$$

$$-\frac{1}{4}\sqrt{5} - \frac{1}{4} + \frac{1}{4}I\sqrt{2}\sqrt{5-\sqrt{5}},$$

$$-\frac{1}{4}\sqrt{5} - \frac{1}{4} - \frac{1}{4}I\sqrt{2}\sqrt{5-\sqrt{5}}, \frac{1}{4}\sqrt{5} - \frac{1}{4} - \frac{1}{4}I\sqrt{2}\sqrt{5+\sqrt{5}}$$

$$]$$

5.4   Procedure Options and the Description Field

Options

A procedure may have one or more options. You may specify options by using the options clause of a procedure definition.

> options *O1*, *O2*, ..., *Om*;

You may use any symbol as an option but the following options have special meanings.

The remember and system Options   When you invoke a procedure with the remember option, Maple stores the result of the invocation in the *remember table* associated with the procedure. Whenever you invoke the procedure, Maple checks whether you have previously called the procedure with the same parameters. If so, Maple retrieves the previously calculated result from the remember table rather than executing the procedure again.

```
> fib := proc(n::nonnegint)
>    option remember;
>    fib(n-1) + fib(n-2);
> end;
```

$$fib := \mathbf{proc}(n:: nonnegint)$$

$$\mathbf{option}\ remember;$$

$$fib(n - 1) + fib(n - 2)$$

$$\mathbf{end}$$

You may place entries in the remember table of a procedure by direct assignment; this method also works for procedures without the remember option.

```
> fib(0) := 0;
```

$$fib(0) := 0$$

```
> fib(1) := 1;
```

$$fib(1) := 1$$

The following is the fib procedure's remember table.

$$table([$$
$$0 = 0$$
$$1 = 1$$
$$])$$

Since fib has the remember option, invoking it places new values in its remember table.

```
> fib(9);
```

$$34$$

Below is the new remember table.

$$table([$$
$$0 = 0$$
$$4 = 3$$
$$8 = 21$$
$$1 = 1$$
$$5 = 5$$
$$9 = 34$$
$$2 = 1$$
$$6 = 8$$
$$3 = 2$$
$$7 = 13$$
$$])$$

The use of remember tables can drastically improve the efficiency of recursively defined procedures.

The system option allows Maple to remove entries from a procedure's remember table. Such selective amnesia occurs during garbage collection, an important part of Maple's memory management scheme. Most importantly, do *not* use the system option with procedures like fib above that depend on entries in their remember table to terminate their execution. See *Remember Tables* on page 68 for more details and examples of remember tables.

The operator and arrow Options   The operator option allows Maple to make additional simplifications to the procedure, and the arrow option indicates that the pretty-printer should display the procedure using the arrow notation.

```
> proc(x)
>     option operator, arrow;
>     x^2;
> end;
```

$$x \to x^2$$

Mapping Notation on page 182 describes procedures using the arrow notation.

The Copyright Option   Maple considers any option that begins with the word *Copyright* to be a Copyright option. Maple does not print the body of a procedure with a Copyright option unless the interface variable verboseproc is at least 2.

```
> f := proc(expr::anything, x::name)
>     option 'Copyright 1684 by G. W. Leibniz';
>     Diff(expr, x);
> end;
```

$$f := \mathbf{proc}(\textit{expr::anything, x::name}) \ldots \mathbf{end}$$

The builtin Option   Maple has two main classes of procedures: those which are part of the Maple kernel, and those which the Maple language itself defines. The builtin option indicates the kernel procedures. You can see this when you fully evaluate a built-in procedure.

```
> eval(type);
```

$$\mathbf{proc}() \ \mathbf{option} \ \textit{builtin}; \ 174 \ \mathbf{end}$$

Each built-in procedure is uniquely identified by a number. Of course, you cannot create built-in procedures of your own.

The Description Field

The last part of the procedure header is the `description` field. It must appear after any `local` clause, `global` clause, or `options` clause, and before the body of the procedure. It takes the following form.

> description *string* ;

The description field has no effect on the execution of the procedure. Its use is for documentation purposes. Unlike a comment, which Maple discards when you read in a procedure, the description field provides a way to attach a one line comment to a procedure.

```
> f := proc(x)
>     description 'computes the square of x';
>     x^2; # compute x^2
> end:
> print(f);
```

$$\textbf{proc}(x)$$

$$\textbf{description } \textit{'computes the square of } x';$$

$$x^2$$

$$\textbf{end}$$

Also, Maple prints the description field even if it does not print the body of a procedure due to a `Copyright` option.

```
> f := proc(x)
>     option 'Copyrighted ?';
>     description 'computes the square of x';
>     x^2; # compute x^2
> end:
> print(f);
```

$$\textbf{proc}(x) \textbf{ description } \textit{'computes the square of } x' \ldots \textbf{end}$$

5.5 The Value Returned by a Procedure

When you invoke a procedure, the value that Maple returns is normally the value of the last statement in the statement sequence of the body of the procedure. Three other types of returns from procedures are a return through a parameter, an *explicit* return, and an *error* return.

Assigning Values to Parameters

Sometimes you may want to write a procedure that returns a value through a parameter. Consider writing a boolean procedure, MEMBER, which determines whether a list L contains an expression x. Moreover, if you call MEMBER with a third argument, p, then MEMBER should assign the position of x in L to p.

```
> MEMBER := proc(x::anything, L::list, p::evaln) local i;
>     for i to nops(L) do
>         if x=L[i] then
>             if nargs>2 then p := i fi;
>             RETURN(true)
>         fi;
>     od;
>     false
> end:
```

If you call MEMBER with two arguments, then nargs is two, so the body of MEMBER does not refer to the formal parameter, p. Therefore, Maple does not complain about a missing parameter.

```
> MEMBER( x, [a,b,c,d] );
```

$$false$$

If you call MEMBER with three arguments, then the type declaration p::evaln ensures that Maple evaluates the third actual parameter to a name[2] rather than using full evaluation.

```
> q := 78;
```

$$q := 78$$

```
> MEMBER( c, [a,b,c,d], q );
```

$$true$$

```
> q;
```

$$3$$

Maple evaluates parameters only once. This means that you cannot use formal parameters freely like local variables within a procedure body. *Once you have made an assignment to a parameter you should not refer to that parameter again.* The only legitimate purpose for assigning to a parameter is so that on return from the procedure the corresponding actual parameter

---

[2]If the third parameter has not been declared evaln, then you should enclose the name q in single quotes ('q') to ensure that the name and not the value of q is passed to the procedure.

has an assigned value. The following procedure assigns the value -13 to its parameter, then returns the name of that parameter.

```
> f := proc(x::evaln)
>     x := -13;
>     x;
> end:
> f(q);
```

$$q$$

The value of q is now -13.

```
> q;
```

$$-13$$

The count procedure below is a more complicated illustration of this phenomenon. count should determine whether a product of factors, p, contains an expression, x. If p contains x, then count should return the number of factors that contain x in the third parameter, n.

```
> count := proc(p::'*', x::name, n::evaln)
>     local f;
>     n := 0;
>     for f in p do
>         if has(f,x) then n := n+1 fi;
>     od;
>     evalb( n>0 );
> end:
```

The count procedure does not work as intended.

```
> count(2*x^2*exp(x)*y, x, m);
```

$$-m < 0$$

The value of the formal parameter n inside the procedure is always m, the actual parameter that Maple determines once and for all when you invoke the procedure. Thus, when execution reaches the evalb statement, the value of n is the name m, and not the value of m. Worse yet, the n:=n+1 statement assigns to m the *name* m+1, as you can see if you evaluate m one level.

```
> eval(m, 1);
```

$$m + 1$$

The m in the above result also has the value m+1.

```
> eval(m, 2);
```

$$m + 2$$

Thus, if you were to evaluate m fully, you would throw Maple into an infinite loop.

A general solution to this type of problem is to use local variables and to view the assignment to a parameter as an operation which takes place just before returning from the procedure.

```
> count := proc(p::'*', x::name, n::evaln)
>     local f, m;
>     m := 0;
>     for f in p do
>         if has(f,x) then m := m + 1 fi;
>     od;
>     n := m;
>     evalb( m>0 );
> end:
```

The new version of count works as intended.

```
> count(2*x^2*exp(x)*y, x, m);
```

$$true$$

```
> m;
```

$$2$$

Explicit Returns

An *explicit return* occurs when you invoke the RETURN command, which has the following syntax.

```
RETURN( sequence );
```

The RETURN command causes an immediate return from the procedure and value of the *sequence* becomes the the value of the procedure invocation.

For example, the following procedure computes the first position i of a value x in a list of values L. If x is not in the list L, the procedure returns 0.

```
> POSITION := proc(x::anything, L::list)
>     local i;
>     for i to nops(L) do
>         if x=L[i] then RETURN(i) fi;
>     od;
>     0;
> end:
```

In most applications of the RETURN command, you will use it to return only a single expression. Returning a sequence, however, including the empty sequence, is quite legitimate. For example, the GCD procedure below computes the greatest common divisor g of two integers a and b. It returns the sequence $g, a/g, b/g$. GCD must treat the case $a = b = 0$ separately because that makes g zero.

```
> GCD := proc(a::integer, b::integer)
>     local g;
>     if a=0 and b=0 then RETURN(0,0,0) fi;
>     g := igcd(a,b);
>     g, iquo(a,g), iquo(b,g);
> end:
> GCD(0,0);
```

$$0, 0, 0$$

```
> GCD(12,8);
```

$$4, 3, 2$$

Of course, instead of returning a sequence, you may also return a list or a set of values.

Error Returns

An *error return* occurs when you invoke the ERROR command which has the following syntax.

ERROR(*sequence*);

The ERROR command normally causes an immediate exit from the current procedure to the Maple session. Maple prints the following error message.

Error, (in *procname*), *sequence*

Here *sequence* is the argument sequence of the ERROR command and *procname* is the name of the procedure in which the error happened. If the procedure does not have a name, Maple says Error, (in unknown) ....

A common use of the ERROR command occurs when you need to check that actual parameters are of the correct type, but parameter declarations are not sufficient for the job. The pairup procedure below takes a list L of the form $[x_1, y_1, x_2, y_2, \ldots, x_n, y_x]$ as input and creates from it a listlist of the form $[[x_1, y_1], [x_2, y_2], \ldots, [x_n, y_n]]$. A simple type check cannot determine if the list L has an even number of elements, so you need to check that explicitly.

```
> pairup := proc(L::list)
>     local i, n;
```

```
>     n := nops(L);
>     if irem(n,2)=1 then
>         ERROR( "L must have an even number of entries" );
>     fi;
>     [seq( [L[2*i-1],L[2*i]], i=1..n/2 )];
> end:
> pairup([1, 2, 3, 4, 5]);
```

Error, (in pairup)
L must have an even number of entries

```
> pairup([1, 2, 3, 4, 5, 6]);
```

$$[[1, 2], [3, 4], [5, 6]]$$

Trapping Errors

The global variable lasterror stores the value of the latest error. The pairup procedure in the previous section caused an error.

```
> lasterror;
```

$$\textit{'L must have an even number of entries'}$$

You can trap errors with the traperror command. The traperror command evaluates its arguments; if no error occurs, it simply returns the evaluated arguments.

```
> x := 0:
> result := traperror( 1/(x+1) );
```

$$result := 1$$

If an error occurs while Maple is evaluating the arguments, then traperror returns the corresponding error.

```
> result := traperror(1/x);
```

$$result := \textit{'division by zero'}$$

Now the value of lasterror has changed.

```
> lasterror;
```

$$\textit{'division by zero'}$$

You can test if an error occurred by comparing the result of traperror with lasterror. If an error did occur, then lasterror and the value which traperror returns are identical.

```
> evalb( result = lasterror );
```

$$\textit{true}$$

A very useful application of the `traperror` and `ERROR` commands is to abort an expensive computation as quickly and cleanly as possible. For example, suppose you are trying to compute an integral using one of several methods, and in the middle of applying the first method you determine that it will not succeed. You would like to abort that method and go on to try another method. The code that tries the different methods might look like the following.

```
> result := traperror( MethodA(f,x) );
> if result=lasterror then   # An error occurred
>    if lasterror=FAIL then  # Method A failed, try Method B
>        result := MethodB(f,x);
>    else # some other kind of error occurred
>        ERROR(lasterror); # propagate that error
>    fi
> else # Method A succeeded
>    RETURN(result);
> fi;
```

`MethodA` can abort its computation at any time by executing the command `ERROR(FAIL)`.

Returning Unevaluated

Maple often uses a particular form of return as a *fail return*, in the sense that it cannot carry out the computation and so returns the unevaluated function invocation as the result. The procedure `MAX`, below, calculates the maximum of two numbers, x and y.

```
> MAX := proc(x,y) if x>y then x else y fi end:
```

The above version of `MAX` is unacceptable for a symbolic computation system because it insists on its arguments being numerical values so that Maple can determine if $x > y$.

```
> MAX(3.2, 2);
```

$$3.2$$

```
> MAX(x, 2*y);
```

```
Error, (in MAX) cannot evaluate boolean
```

The absence of symbolic capabilities in `MAX` causes problems when you try to plot expressions involving `MAX`.

```
> plot( MAX(x, 1/x), x=1/2..2 );
```

```
Error, (in MAX) cannot evaluate boolean
```

The error occurs because Maple evaluates MAX(x, 1/x) before invoking the plot command.

The solution is to make MAX return unevaluated when its parameters, x and y, are not numeric. That is, in such cases MAX should return 'MAX'(x,y).

```
> MAX := proc(x, y)
>     if type(x, numeric) and type(y, numeric) then
>         if x>y then x else y fi;;
>     else
>         'MAX'(x,y);
>     fi;
> end:
```

The new version of MAX handles both numeric and non-numeric input.

```
> MAX(3.2, 2);
```

$$3.2$$

```
> MAX(x, 2*y);
```

$$MAX(x,\ 2\,y)$$

```
> plot( MAX(x, 1/x), x=1/2..2 );;
```

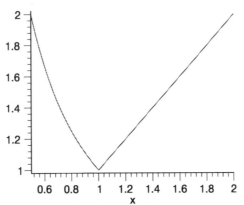

You can improve MAX so that it can find the maximum of any number of arguments. Inside a procedure, args is the sequence of actual parameters, nargs is the number of actual parameters, and procname is the name of the procedure.

```
> MAX := proc()
>     local m, i;
```

```
>     m := -infinity;
>     for i in (args) do
>         if not type(i, numeric) then
>             RETURN('procname'(args));
>         fi;
>         if i>m then m := i fi;
>     od;
>     m;
> end:
> MAX(3,1,4);
```

$$4$$

```
> MAX(3,x,1,4);
```

$$MAX(3, x, 1, 4)$$

The sin function and the int integration command essentially work like the MAX procedure above. If Maple can compute the result, it returns it; otherwise, sin and int return unevaluated.

Exercise

1. Improve the MAX procedure above so that MAX(3,x,1,4) returns MAX(x,4); that is, the procedure returns the maximum numerical value along with all non-numerical values.

5.6 The Procedure Object

This section describes the procedure object, its type and operands, its special evaluation rule, and how to save it to a file and retrieve it again.

Last Name Evaluation

Maple evaluates ordinary expressions in a *full recursive evaluation* mode. All future references to a name that you assign a value return the computed value instead of the name.

```
> f := g;
```

$$f := g$$

```
> g := h;
```

$$g := h$$

```
> h := x^2;
```

$$h := x^2$$

Now f evaluates to x^2.

```
> f;
```

$$x^2$$

Names of procedures, arrays, and tables are exceptions. For such names, Maple uses a *last-name evaluation* model. This model avoids printing all the details forming the procedure definition.

```
> F := G;
```

$$F := G$$

```
> G := H;
```

$$G := H$$

```
> H := proc(x) x^2 end;
```

$$H := \mathbf{proc}(x)\, x^2 \ \mathbf{end}$$

Now F evaluates to H because H is the last name before the actual procedure.

```
> F;
```

$$H$$

You can use the eval command to evaluate a procedure fully.

```
> eval(F);
```

$$\mathbf{proc}(x)\, x^2 \ \mathbf{end}$$

See also *Evaluation Rules* on page 42.

The Type and Operands of a Procedure

Maple recognizes all procedures (including those created using the mapping notation) as being of type procedure, as are any names that you give to procedures.

```
> type(F,name);
```

$$true$$

```
> type(F,procedure);
```

$$true$$

```
> type(eval(F),procedure);
```

true

Thus, you can use the following test to ensure that *F* is the name of a procedure.

```
> if type(F, name) and type(F, procedure) then ... fi
```

A procedure has six operands:

1. the sequence of formal parameters
2. the sequence of local variables
3. the sequence of options
4. the remember table
5. the description string
6. the sequence of global variables

As an example of the structure of a procedure, consider the following.

```
> f := proc(x::name, n::posint)
>     local i;
>     global y;
>     option Copyright;
>     description "a summation";
>     sum( x[i] + y[i], i=1..n );
> end:
```

Place an entry in the procedure's remember table.

```
> f(t,3) := 12;
```

$$f(t, 3) := 12$$

You can see the various parts of *f* below.

The name of the procedure:

```
> f;
```

$$f$$

The procedure itself:

```
> eval(f);
```

$$\textbf{proc}(x::name, \; n::posint)$$

$$\textbf{description } \text{``a summation''}$$

$$\cdots$$

$$\textbf{end}$$

The formal parameters:

```
> op(1, eval(f));
```

$$x::name, \ n::posint$$

The local variables:

```
> op(2, eval(f));
```

$$i$$

The options:

```
> op(3, eval(f));
```

$$Copyright$$

The remember table:

```
> op(4, eval(f));
```

$$table([$$
$$(t, 3) = 12$$
$$])$$

The description:

```
> op(5, eval(f));
```

$$\text{"a summation"}$$

The global variables:

```
> op(6, eval(f));
```

$$y$$

The body of a procedure is *not* one of its operands, so you cannot gain access to the body with the op command. If you need to manipulate the body of a procedure, see ?CompSeq or ?codegen.

Saving and Retrieving Procedures

While you develop a new procedure, you can save your work by saving the whole worksheet; when you are satisfied that your procedure works the way you intend, you may want to save it in a .m file. Such files use Maple's internal format which allows Maple to efficiently retrieve these stored objects.

```
> CMAX := proc(x::complex(numeric), y::complex(numeric))
>     if abs(x)>abs(y) then
>         x;
>     else
>         y;
>     fi;
> end:
```

Use the save command to save procedures in the same manner you save any other Maple object.

```
> save CMAX, "CMAX.m";
```

The read command retrieves the objects stored in a .m file.

```
> read "CMAX.m";
```

Some Maple users prefer to write Maple procedures with their favorite text editor. You can also use the read command to read in data from such files. Maple executes each line in the file as if you had typed it directly into your session.

If you make a number of related procedures, you may want to save them as a Maple package. Making a package allows you to load the procedures using the with command. See *Writing Your Own Packages* on page 98.

5.7   Explorations

The purpose of the exercises in this section is to deepen your understanding of how Maple procedures work. In some cases you may wish to study the on-line help pages for the various Maple commands that you will need.

Exercises

1. Implement the function $f(x) = (\sqrt{1-x^2})^3 - 1$, first as a procedure, then using the mapping notation. Compute $f(1/2)$ and $f(0.5)$ and comment on the different results. Use the D operator to compute f', and then compute $f'(0)$.

2. Write a procedure, SPLIT, which on input of a product f and a variable x returns a list of two values. The first item in the list should be the product of the factors in f that are independent of x, and the second item should be the product of the factors that have an x in them. *Hint:* You may want to use the has, select, and remove commands.

3. The following program tries to compute $1 - x^{|a|}$.

```
> f := proc(a::integer, x::anything)
```

```
>        if a<0 then a := -a fi;
>        1-x^a;
> end:
```

What is wrong with this procedure? You may want to use the Maple debugger to isolate the error. See chapter 6.

4. ab/g gives the least common multiple of two integers, a and b, where g is the greatest common divisor of a and b. For example, the least common multiple of 4 and 6 is 12. Write a Maple procedure, LCM, which takes as input $n > 0$ integers $a_1, a_2, \ldots, a_n$ and computes their least common multiple. By convention, the least common multiple of zero and any other number is zero.

5. The following recurrence relation defines the Chebyshev polynomials of the first kind, $T_n(x)$.

$$T_0(x) = 1, \qquad T_1(x) = x, \qquad T_n(x) = 2xT_{n-1}(x) - T_{n-2}(x)$$

The following procedure computes $T_n(x)$ in a loop for any given integer n.

```
> T := proc(n::integer, x)
>    local t1, tn, t;
>    t1 := 1; tn := x;
>    for i from 2 to n do
>        t := expand(2*x*tn - t1);
>        t1 := tn; tn := t;
>    od;
>    tn;
> end:
```

The procedure has several errors. Which variables should have been declared local? What happens if n is zero or negative? Identify and correct all errors, using the Maple debugger where appropriate. Modify the procedure so that it returns unevaluated if n is a symbolic value.

5.8 Conclusion

In this chapter, you have seen the details of the proc command. You have learned the finer points of the options at your disposal when defining procedures. You have learned about functional operators, unnamed procedures, and procedure simplification.

In addition, you have thoroughly reviewed Maple's evaluation rules which chapter 2 introduced. For example, Maple generally evaluates local variables to one level and global variables fully. Maple evaluates the arguments to a procedure at the time you invoke it. How they are evaluated depends upon the environment in which the call occurs, and in some cases, the types specified within the procedure definition. Once evaluated, Maple substitutes the values into the procedure and then executes it. Maple does no further evaluation on the values which it substituted, unless you specifically use a command such as eval. This rule makes it impractical to use parameters to store temporary results, as you would use local variables.

Although chapters 1 and 2 discussed type declarations, this chapter reviewed these declarations in full. Type declarations are particularly useful as a means of stating the intended purpose of your procedures and as a convenient means of supplying error messages to any user who might call them with inappropriate values.

This chapter concludes the formal review of the Maple language which began in chapter 4. The remaining chapters deal with specific areas of Maple programming. For example, chapter 6 discusses the Maple debugger, chapter 7 introduces you to the details of numerical programming in Maple, and chapter 8 shows how to extend Maple's extensive plotting facilities to suit your needs.

6 Debugging Maple Programs

A program, whether developed in Maple or any other language, often works incorrectly when first tested due to logic errors in the design or errors introduced during implementation. Many errors can be very subtle and hard to find by experimentation and visual inspection of the program alone. Maple provides a debugger to help you find these errors.

The Maple debugger lets you stop execution within a Maple procedure, inspect and/or modify the values of local and global variables, and continue execution, either to completion, or one statement or block at a time. You can stop execution when Maple reaches a particular statement, when it assigns some value to a particular local or global variable, or when a particular error occurs. These facilities let you watch the inner workings of your program to determine why it is not doing what you expect.

6.1 A Tutorial Example

This section describes how to use the debugger to debug a Maple procedure and uses the following code as a case in point.

```
> sieve := proc(n::integer)
>     local i,k,flags,count,twice_i;
>     count := 0;
>     for i from 2 to n do flags[i] := true od;
>     for i from 2 to n do
>         if flags[i] then
>             twice_i := 2*i;
>             for k from twice_i by i to n do
>                 flags[k] = false;
```

```
>              od;
>              count := count+1
>         fi;
>     od;
>     count;
> end:
```

This procedure implements the Sieve of Eratosthenes. Given a parameter n, this procedure returns a count of all the prime numbers less than n inclusive. While debugging this procedure, sieve, you should find several errors.

Many debugger commands need to refer to statements within the procedures you are debugging. Statement numbers make such references. A statement does not necessarily correspond to a line of text in a source file; one source line could correspond to part of a statement, or to more than one statement. The showstat command, described in more detail in *Displaying the Statements of a Procedure* on page 218, displays Maple procedures with statement numbers, and illustrates the distinction between source lines and statements.

```
> showstat(sieve);

sieve := proc(n::integer)
local i, k, flags, count, twice_i;
    1      count := 0;
    2      for i from 2 to n do
    3        flags[i] := true
           od;
    4      for i from 2 to n do
    5        if flags[i] then
    6          twice_i := 2*i;
    7          for k from twice_i by i to n do
    8            flags[k] = false
             od;
    9          count := count+1
           fi
         od;
   10      count
end
```

To invoke the debugger you must start execution of the procedure, and execution must be made to stop within the procedure. To execute a Maple procedure, call it from a Maple command at the top level, or call it from within another procedure. The simplest way to cause execution to stop within the procedure is to set a *breakpoint* in the procedure. To do this, use the stopat command.

```
> stopat(sieve);
```

[*sieve*]

This sets a breakpoint before the first statement in procedure sieve. The stopat command also returns a list of all procedures containing breakpoints (sieve, in this case). When you subsequently execute sieve, Maple will stop before executing the first statement within it. When execution stops, the debugger prompt appears. The following example demonstrates an initial execution of sieve.

```
> sieve(10);

sieve:
    1*    count := 0;


DBG>
```

Preceding the debugger prompt ("DBG>" in the examples) are several pieces of information.

1. The previously computed result (this particular execution has stopped before making any computations, so no result appears).

2. The name of the procedure in which execution has stopped (sieve, in this example).

3. The statement number before which execution stopped (1 in this example). An asterisk (*) or question mark (?) may follow the statement number indicating that a breakpoint or conditional breakpoint has been set before the statement.

4. Rather than writing out all the statements contained at deeper levels within a compound statement such as if or do, Maple displays the body of such statements as "...".

While at the debugger prompt, you can evaluate Maple expressions, and invoke debugger commands. Maple evaluates expressions in the context of the stopped procedure. You have access to exactly the same procedure parameters, local, global, and environment variables as the stopped procedure has. You called sieve with parameter 10, so now the formal parameter n has the value 10.

```
DBG> n

10
sieve:
```

```
1*    count := 0;
```

Notice that for each expression that Maple evaluates, it displays the result, followed by the name of the stopped procedure, the statement number, the statement, and a new debugger prompt.

Debugger commands control execution once the debugger is active. The most commonly used debugger command is next, which executes the displayed statement, and causes execution to stop before the next statement at the same (or shallower) nesting level.

```
DBG> next

0
sieve:
    2    for i from 2 to n do
             ...
         od;
```

The 0 in the first line of the output represents the result of the assignment count := 0. Also, no "*" appears next to the statement number, because there is no breakpoint before statement 2. The debugger does not show the body of the for loop, which itself consists of statements (with their own statement numbers), until execution actually stops within the body.

Executing the next command again results in

```
DBG> next

true
sieve:
    4    for i from 2 to n do
             ...
         od;
```

Execution now stops before statement 4. Statement 3 (the body of the previous for loop) is at a deeper nesting level, and therefore the next command skipped over it. Although it executed (n−1 times), execution did not stop within the loop. However, the debugger displays the last result computed in the loop (the assignment of the value true to flags[10]).

To step into a nested control structure, use the step debugger command.

```
DBG> step

true
sieve:
    5    if flags[i] then
             ...
```

```
            fi
DBG> step

    true
    sieve:
       6          twice_i := 2*i;
```

If you use the `step` debugger command when the statement to execute is not a deeper structured statement, it has the same effect as the `next` debugger command.

```
DBG> step

    4
    sieve:
       7          for k from twice_i by i to n do
                     ...
                  od;
```

At this point, using the `showstat` debugger command to display the entire procedure might be worthwhile.

```
DBG> showstat
sieve := proc(n::integer)
local i, k, flags, count, twice_i;
    1*    count := 0;
    2     for i from 2 to n do
    3       flags[i] := true
          od;
    4     for i from 2 to n do
    5       if flags[i] then
    6         twice_i := 2*i;
    7 !       for k from twice_i by i to n do
    8           flags[k] = false
            od;
    9         count := count+1
          fi
        od;
   10     count
end
```

The `showstat` debugger command is similar to the ordinary `showstat` command, except that it only works within the debugger, and displays the currently stopped procedure if you do not name a procedure. An exclamation point (!) following the statement number indicates the statement at which you are stopped.

Step into the innermost loop.

```
DBG> step

4
sieve:
   8            flags[k] = false
```

A related debugger command, list, displays only the previous five statements, the current statement, and the next statement, to quickly provide some idea of where the procedure has stopped.

```
DBG> list

sieve := proc(n::integer)
local i, k, flags, count, twice_i;
      . . .
   3      flags[i] := true
        od;
   4    for i from 2 to n do
   5      if flags[i] then
   6        twice_i := 2*i;
   7        for k from twice_i by i to n do
   8 !         flags[k] = false
          od;
   9        count := count+1
        fi
      od;
      . . .

end
```

Use the outfrom debugger command to finish execution at the current nesting level or deeper, and then to stop once again when execution reaches a statement at a shallower nesting level.

```
DBG> outfrom

true = false
sieve:
   9            count := count+1

DBG> outfrom

1
sieve:
   5      if flags[i] then
            . . .
        fi
```

The cont debugger command continues execution, either until it terminates normally, or until it encounters another breakpoint.

```
DBG> cont
```

                                    91

You can now see that the procedure does not give the expected output. Although you may find the reason obvious from the previous debugger command examples, in a real-life case this is often not so. Therefore, pretend not to recognize the problem, and continue to use the debugger. First, use the unstopat command to remove the breakpoint from sieve.

```
> unstopat(sieve);
```

                                    []

The procedure sieve keeps track of the changing result in the variable count. A logical place to look then is wherever Maple modifies count. The easiest way to do this is using a *watchpoint*. A watchpoint invokes the debugger whenever Maple modifies a watched variable. Use the stopwhen command to set watchpoints. Here you want to stop when Maple modifies the variable count in the procedure sieve.

```
> stopwhen([sieve,count]);
```

                         [[*sieve, count*]]

The stopwhen command returns a list of all the variables currently being watched.

Execute the procedure sieve again.

```
> sieve(10);

count := 0
sieve:
    2    for i from 2 to n do
             ...
         od;
```

When execution stops because Maple has modified count, the debugger displays the assignment statement count := 0. The debugger then, as usual, displays the name of the procedure and the next statement in the procedure. Note that execution stops *after* Maple has already assigned a value to count.

Clearly, this first assignment to count is doing the right thing, so use the cont debugger command to continue execution.

```
DBG> cont

count := 1
sieve:
    5        if flags[i] then
```

```
        . . .
        fi
```

If you do not look carefully enough, this also looks correct, so continue execution.

```
DBG> cont

count := 2*l
sieve:
   5        if flags[i] then
        . . .
        fi
```

This output is suspicious, because Maple would have simplified 2*1, but notice that it has printed 2*l instead. Looking at the source text for the procedure, reveals that the letter "l" was typed instead of the number "1". There is no point continuing execution any further, so use the quit debugger command to exit the debugger.

```
DBG> quit
```

After correcting the source text and reading it back into Maple, turn off the watchpoint and try the example again.

```
> unstopwhen();
```

$$[]$$

```
> sieve(10);
```

$$9$$

This is still incorrect. There are four primes less than 10, namely 2, 3, 5, and 7. Therefore, invoke the debugger once more, stepping into the innermost parts of the procedure to investigate. Not wanting to start at the very beginning of the procedure, set the breakpoint at statement 6.

```
> stopat(sieve,6);
```

$$[sieve]$$

```
> sieve(10);

true
sieve:
   6*       twice_i := 2*i;

DBG> step

4
sieve:
```

```
7           for k from twice_i by i to n do
            ...
        od;
```

```
DBG> step
```

```
4
sieve:
   8             flags[k] = false
```

```
DBG> step
```

```
true = false
sieve:
   8             flags[k] = false
```

The last step shows the error. The previously computed result should have been `false` (from the assignment of `flags[k]` to the value `false`), but instead was `true = false`. What should have been an assignment was really an equation instead, and therefore Maple did not set `flags[k]` to `false` at all. Again, exit the debugger and correct the source text.

```
DBG> quit
```

Note that when execution of a Maple command is interrupted for any reason, including quitting from the debugger, Maple displays the message "Interrupted".

This is the corrected procedure.

```
> sieve := proc(n::integer)
> local i,k,flags,count,twice_i;
>     count := 0;
>     for i from 2 to n do flags[i] := true od;
>     for i from 2 to n do
>         if flags[i] then
>             twice_i := 2*i;
>             for k from twice_i by i to n do
>                 flags[k] := false;
>             od;
>             count := count+1
>         fi
>     od;
>     count
> end:
```

Now `sieve` gives the correct answer.

```
> sieve(10);
```

4

6.2 Invoking the Debugger

You invoke the Maple debugger by means of breakpoints, watchpoints, or error watchpoints. This section describes how to set such break- and watchpoints. Before you can set any breakpoints, you need to know the statement numbers of the procedure(s) you want to examine.

Displaying the Statements of a Procedure

Use the showstat command to display the statements of a procedure along with statement numbers as determined by the debugger. The showstat command can be used with the following syntax.

```
showstat( procedure )
```

Here *procedure* is the name of the procedure to be displayed. The showstat command marks unconditional breakpoints (see *Breakpoints* on page 219) with a "*" and conditional breakpoints with a "?".

You can also use the showstat command to display a single statement or a range of statements.

```
showstat( procedure, number )
showstat( procedure, range )
```

In these cases the statements that are not displayed are shown as "...". The procedure name, its parameters, and its local and global variables are always displayed.

You can also display the statements of a procedure from within the debugger. The showstat debugger command has the following syntax.

```
showstat procedure number_or_range
```

The arguments are the same as for the ordinary showstat command, except that you can also omit the *procedure*. In that case, the debugger shows the specified statements (all statements if you do not specify any) of the currently stopped procedure. When displaying the currently stopped procedure, the showstat debugger command displays an exclamation point (!) next to the statement number at which the debugger has stopped execution.

A number of commands exist in slightly different forms inside and outside the debugger. The showstat command is one example. Usually the difference is that inside the debugger you do not have to specify the name of the procedure. Outside the debugger, you must enclose the arguments of a command in parentheses and separate them by commas.

Breakpoints

Use *breakpoints* to invoke the debugger at a specific statement in a procedure. You can set breakpoints in Maple procedures using the stopat command. Use the following method for invoking the stopat command.

stopat(*procedureName*, *statementNumber*, *condition*)

Here *procedureName* is the name of the procedure in which to set the breakpoint, and *statementNumber* is the statement number within the procedure before which to set the breakpoint. If you omit the *statementNumber*, the debugger sets a breakpoint before statement 1 of the procedure (that is, execution will stop as soon you invoke as the procedure). Unconditional breakpoints set with stopat are marked by a "*" when showstat displays the procedure.

The optional *condition* argument specifies a boolean condition which must be true for execution to stop. This condition can refer to any global variables, local variables, or parameters of the procedure. Conditional breakpoints set with stopat are marked by a "?" when showstat displays the procedure.

You can also set breakpoints from within the debugger. The stopat debugger command has the following syntax.

stopat *procedureName* *statementNumber* *condition*

The arguments are the same as for the ordinary stopat command, except that you can also omit the procedure name. In that case, the debugger sets the breakpoint at the specified statement (the first statement if you do not specify any) of the currently stopped procedure.

Note that stopat sets a breakpoint *before* the specified statement. When Maple encounters a breakpoint, execution stops and Maple engages the debugger before the statement. This means that it is impossible to set a breakpoint after the last statement in a statement sequence (that is, at the end of a loop body, if statement body, or procedure).

If two identical procedures exist, they may or may not share breakpoints depending on how you created them. If you wrote the procedures individually, with textually identical procedure bodies, then they do *not* share breakpoints. If you created one by assigning it the body of the other, then they *will* share breakpoints.

```
> f := proc(x) x^2 end:
> g := proc(x) x^2 end:
> h := op(g):
```

> stopat(g);

$$[g, h]$$

Use the unstopat command to clear breakpoints. Invoke the unstopat command as follows.

```
unstopat( procedureName, statementNumber )
```

Here *procedureName* is the name of the procedure from which to clear a breakpoint, and *statementNumber* is the statement number within the procedure. If you omit the *statementNumber*, then unstopat clears *all* breakpoints in the procedure.

You can remove breakpoints from inside the debugger. The unstopat debugger command has the following syntax.

```
unstopat procedureName statementNumber
```

The arguments are the same as for the ordinary unstopat command, except that you can also omit the *procedureName*. This clears the breakpoint from the specified statement (all breakpoints if you do not specify any) of the currently stopped procedure.

Explicit Breakpoints   You can insert an explicit breakpoint into the source text of a procedure by writing a call to the DEBUG command.

```
DEBUG()
```

If you call the DEBUG command with no arguments, then execution stops at the statement following the DEBUG command and the debugger is invoked.

If the argument of the DEBUG command is a boolean expression,

```
DEBUG( boolean )
```

then execution stops only if the *boolean* expression evaluates to true. If the *boolean* expression evaluates to false or FAIL, then the DEBUG command is ignored.

If the argument of the DEBUG command is anything but a boolean expression,

```
DEBUG( not boolean )
```

then the debugger prints the value of the argument instead of the last result when execution stops at the following statement.

```
> f := proc(x)
>    DEBUG("my breakpoint, current value of x:",x);
>    x^2
> end:
> f(3);

"my breakpoint, current value of x:"
3
f:
    2    x^2
```

The showstat command does *not* mark explicit breakpoints with a "∗" or a "?".

```
DBG> showstat

f := proc(x)
    1    DEBUG("my breakpoint, current value of x:",x);
    2 !  x^2
end
```

The unstopat command *cannot* remove explicit breakpoints.

```
DBG> unstopat

[f]
f:
    2    x^2

DBG> showstat

f := proc(x)
    1    DEBUG("my breakpoint, current value of x:",x);
    2 !  x^2
end

DBG> quit
```

Breakpoints that you insert using stopat appear as a call to DEBUG if you display the procedure using print or lprint.

```
> f := proc(x) x^2 end:
> stopat(f);
```

$$[f]$$

```
> print(f);
```

$$\mathbf{proc}(x)\,\mathrm{DEBUG}();\ x^2\ \mathbf{end}$$

Watchpoints

Watchpoints monitor local or global variables, and stop execution whenever you assign a value to them. Watchpoints are a useful alternative to breakpoints when you want execution to stop based on *what* is happening instead of *where* it is happening.

You may set watchpoints using the stopwhen command. Invoke the stopwhen command in one of the following two ways.

```
stopwhen( globalVariableName )
stopwhen( [procedureName, variableName] )
```

The first form specifies that the debugger should be invoked whenever the global variable *globalVariableName* is changed. You can also monitor Maple environment variables, such as Digits, this way.

```
> stopwhen(Digits);
```

$$[Digits]$$

The second form specifies that the debugger should be invoked whenever the (local or global) variable *variableName* is changed in the procedure *procedureName*. In either form (or when you call stopwhen with no arguments), Maple returns a list of the currently set watchpoints.

```
> f := proc(x)
>    local a;
>    x^2;
>    a:=%;
>    sqrt(a);
> end:
> stopwhen([f, a]);
```

$$[Digits, [f, a]]$$

When execution stops because Maple has modified a watched variable, the debugger displays an assignment statement instead of the last computed result (which would be the right-hand side of the assignment statement). The debugger then, as usual, displays the name of the procedure and the next statement in the procedure. Note that execution stops *after* Maple has already assigned a value to the watched variable.

You may also set watchpoints with the stopwhen debugger command. Invoke the stopwhen debugger command from within the debugger in the following manner.

```
stopwhen globalVariableName
stopwhen [procedureName variableName]
```

The arguments are the same as for the ordinary stopwhen command.

Clear watchpoints using the unstopwhen command (or the unstopwhen debugger command). The arguments are the same as for stopwhen. If you do not specify any arguments to unstopwhen, then *all* watchpoints are cleared. Similar to stopwhen, unstopwhen returns a list of all (remaining) watchpoints.

Error Watchpoints

Use error watchpoints to monitor Maple errors. When a watched error occurs, execution stops and the debugger displays the statement in which the error occurred.

Use the stoperror command to set error watchpoints. Invoke the stoperror command as follows.

> stoperror(*"errorMessage"*)

This specifies that the debugger should be invoked whenever the error message *errorMessage* is issued. You may use the name all for *errorMessage* to specify that execution should stop when *any* error message is issued.

The stoperror command returns a list of the currently set error watchpoints; if you call stoperror with no arguments, that is all it does.

```
> stoperror("division by zero");
```

$$[\,`division\ by\ zero`\,]$$

Errors trapped by traperror do not generate an error message and so stoperror cannot catch them. Use the command stoperror(traperror) to invoke the debugger whenever a trapped error occurs.

You may also set error watchpoints from within the debugger. The stoperror debugger command has the following syntax.

> stoperror *errorMessage*

The arguments are the same as for the ordinary stoperror command, except that double quotes are not required around the error message.

In addition to the special names 'all' and 'traperror', the stoperror command understands the following special errors.

- 'interrupted' — stop when execution is interrupted by the user (that is, using *Ctrl-C* or Ctrl-Break).

- 'time expired' — stop when execution times out under a call to the timelimit function. See ?timelimit for details.

- `assertion failed` — stop if an assertion failure occurs. See ?ASSERT for details on assertions.

- `invalid arguments` — stop if a function is passed incorrect or insufficient arguments.

The following errors are considered critical and *cannot* be caught by the debugger.

- `out of memory` — Maple has run out of memory and cannot continue.

- `stack overflow` — Maple has run out of stack space and the current computation has been terminated.

- `object too large` — an attempt was made to allocate a single object too large for Maple's capabilities.

Clear error watchpoints using the unstoperror command. The arguments are the same as for stoperror. If you do not specify any arguments to unstoperror, then *all* error watchpoints are cleared. Similar to stoperror, unstoperror returns a list of all (remaining) error watchpoints.

```
> unstoperror();
```

$$[]$$

When execution stops due to an error, you cannot continue executing. Any of the execution control commands, such as next or step, process the error as if the debugger had not intervened.

This example defines two procedures. The first procedure, f, calculates $1/x$. The other, g, calls f but traps the "division by zero" error that happens when $x = 0$.

```
> f := proc(x) 1/x end:
> g := proc(x) local r;
>     r := traperror(f(x));
>     if r = lasterror then infinity
>     else r
>     fi
> end:
```

When trying out the procedure you get the reciprocal at $x = 9$.

```
> g(9);
```

$$\frac{1}{9}$$

And at 0 you get ∞ as expected.

```
> g(0);
```

$$\infty$$

The `stoperror` command stops execution when you call `f` directly.

```
> stoperror("division by zero");
```

$$[\text{`division by zero`}]$$

```
> f(0);

Error, division by zero
f:
    1    1/x

DBG> cont

Error, (in f) division by zero
```

The call to `f` from `g` is inside a `traperror`, so the "division by zero" error does not invoke the debugger.

```
> g(0);
```

$$\infty$$

Instead, try to use `stoperror(traperror)`.

```
> unstoperror("division by zero");
```

$$[]$$

```
> stoperror("traperror");
```

$$[\text{traperror}]$$

Now Maple does not stop at the error in `f`,

```
> f(0);

Error, (in f) division by zero
```

but Maple invokes the debugger when the trapped error occurs.

```
> g(0);

Error, division by zero
f:
    1    1/x

DBG> step

Error, division by zero
1
g:
```

```
    2     if r = lasterror then
              ...
          else
              ...
          fi
DBG> step

Error, division by zero
g:
    3         infinity

DBG> step
```

$$\infty$$

6.3 Examining and Changing the State of the System

When execution stops, you can examine the state of global variables, local variables, and parameters of the stopped procedure. You can also evaluate expressions, determine the stopping point of execution (both on a static and dynamic level), and examine procedures.

The debugger can evaluate any Maple expression and perform assignments to local and global variables. To evaluate an expression, simply type the expression at the debugger prompt.

```
> f := proc(x) x^2 end:
> stopat(f);
```

$$[f]$$

```
> f(10);

f:
    1*    x^2

DBG> sin(3.0)

.1411200081
f:
    1*    x^2

DBG> cont
```

$$100$$

The debugger evaluates any variable names that you use in the expression in the context of the stopped procedure. Names of parameters or local variables evaluate to their current values within the procedure. Names of global variables evaluate to their current values. Environment variables, such as `Digits`, evaluate to their values in the stopped procedure's environment.

If an expression happens to correspond to a debugger command (for example, your procedure has a local variable named `step`), you can still evaluate it by enclosing it in parentheses.

```
> f := proc(step)
>     local i;
>     for i to 10 by step do i^2 od;
> end:
> stopat(f,2);
```

$$[f]$$

```
> f(3);

f:
    2*      i^2

DBG> step

1
f:
    2*      i^2

DBG> (step)

3
f:
    2*      i^2

DBG> quit
```

While execution is halted, you can modify local and global variables using the assignment operator (`:=`). The following example sets a breakpoint in the loop only when the index variable is equal to 5.

```
> sumn := proc(n)
>     local i, sum;
>     sum := 0;
>     for i to n do sum := sum + i od;
> end:
> showstat(sumn);

sumn := proc(n)
local i, sum;
    1       sum := 0;
    2       for i to n do
    3           sum := sum+i
        od
```

```
end

> stopat(sumn,3,i=5);
```

$$[f, sumn]$$

```
> sumn(10);

10
sumn:
   3?      sum := sum+i
```

Reset the index to 3, so that the breakpoint is encountered again later.

```
DBG> i := 3

sumn:
   3?      sum := sum+i

DBG> cont

17
sumn:
   3?      sum := sum+i
```

Now Maple has added together the numbers 1, 2, 3, 4, 3 and 4. Continuing a second time, the procedure finishes by adding to this the numbers 5, 6, 7, 8, 9, and 10.

```
DBG> cont
```

62

Two debugger commands give information about the state of execution. The list debugger command shows you the location within a procedure where execution stopped, and the where debugger command shows you the stack of procedure activations.

Use the list debugger command in the following manner.

```
list procedureName statementNumber
```

The list debugger command is similar to showstat, except in the case where you do not specify any arguments. In that case, list shows the five previous statements, the current statement, and the next statement. This provides some context in the stopped procedure. In other words, it indicates the static position where execution stopped.

The where debugger command shows you the stack of procedure activations. Starting from the top level, it shows you the statement that is executing and the parameters it passed to the called procedure. The where debugger command repeats this for each level of procedure call until it reaches the current statement in the current procedure. In other words, it indicates the dynamic position where execution stopped. The syntax of the where command is as follows.

```
where numLevels
```

This procedure calls the sumn procedure above.

```
> check := proc(i)
>     local p, a, b;
>     p := ithprime(i);
>     a := sumn(p);
>     b := p*(p+1)/2;
>     evalb( a=b );
> end:
```

There is a (conditional) breakpoint in sumn.

```
> showstat(sumn);

sumn := proc(n)
local i, sum;
    1     sum := 0;
    2     for i to n do
    3?        sum := sum+i
          od
end
```

When check calls sumn, the breakpoint invokes the debugger.

```
> check(9);

10
sumn:
    3?        sum := sum+i
```

The where debugger command tells that check was invoked from the top level with argument "9", then check called sumn with argument "23", and now execution is stopped at statement number 3 in sumn.

```
DBG> where

TopLevel: check(9)
        [9]
check: a := sumn(p)
        [23]
sumn:
    3?        sum := sum+i

DBG> cont
```

true

The following example illustrates the use of where when you use it in a recursive function.

```
> fact := proc(x)
>      if x <= 1 then 1
>      else x * fact(x-1) fi;
> end:
> showstat(fact);

fact := proc(x)
    1      if x <= 1 then
    2        1
           else
    3          x*fact(x-1)
           fi
end

> stopat(fact,2);
```

$$[f, \; fact, \; sumn]$$

```
> fact(5);

fact:
    2*      1

DBG> where

TopLevel: fact(5)
        [5]
fact: x*fact(x-1)
        [4]
fact: x*fact(x-1)
        [3]
fact: x*fact(x-1)
        [2]
fact: x*fact(x-1)
        [1]
fact:
    2*      1
```

If you are not interested in the whole history of nested procedure calls, then you can tell where to print out only a certain number of levels.

```
DBG> where 3

fact: x*fact(x-1)
        [2]
fact: x*fact(x-1)
        [1]
fact:
```

```
    2*      1
DBG> quit
```

The `showstop` command (or the `showstop` debugger command) displays a report of all currently set breakpoints, watchpoints, and error watchpoints. Outside the debugger, the `showstop` command takes the form

> showstop();

From within the debugger, use the `showstop` debugger command.

> showstop

Define a procedure.

```
> f := proc(x)
>    local y;
>    if x < 2 then
>        y := x;
>        print(y^2);
>    fi;
>    print(-x);
>    x^3;
> end:
```

Set some breakpoints.

```
> stopat(f):
> stopat(f,2):
> stopat(int);
```

$$[f, \; fact, \; int, \; sumn]$$

Set some watchpoints.

```
> stopwhen(f,y):
> stopwhen(Digits);
```

$$[[f, \; y], \; Digits]$$

Set an error watchpoint.

```
> stoperror("division by zero");
```

$$['division\;by\;zero']$$

The `showstop` command reports all the break- and watchpoints, including those in the earlier procedure f.

```
> showstop();

Breakpoints in:
```

```
       f
       fact
       int
       sumn
Watched variables:
    y in procedure f
    Digits
Watched errors:
    'division by zero'
```

6.4 Controlling Execution

When a breakpoint, watchpoint, or error watchpoint causes Maple to stop executing statements and invoke the debugger, it displays a debugger prompt. While the procedure is stopped you may examine the values of variables or perform other experiments. At that point, you can cause execution to continue in a number of ways.

The examples below use the following two procedures.

```
> f := proc(x)
>     if g(x) < 25 then
>         print("less than five");
>         x^2;
>     fi;
>     x^3;
> end:
> g := proc(x)
>     2*x;
>     x^2;
> end:
> showstat(f);

f := proc(x)
   1     if g(x) < 25 then
   2         print("less than five");
   3         x^2
         fi;
   4     x^3
end

> showstat(g);

g := proc(x)
   1     2*x;
   2     x^2
```

```
end

> stopat(f);
```

$$[f]$$

The quit debugger command exits the debugger and returns to the Maple prompt. The cont debugger command tells the debugger to continue executing statements until a break- or watchpoint is encountered or the procedure terminates normally.

The next debugger command executes the current statement. If the statement is a control structure (an if statement or a loop), the debugger executes any statements within the control structure that it would normally execute and stops execution before the next statement after the control structure. Likewise, if the statement contains calls to procedures, the debugger executes these procedure calls in their entirety before execution stops once more. If there are no further statements after the current statement (for example, if it is the last statement within a branch of an if statement or the last statement in a procedure), then execution does not stop until the debugger encounters another statement (at a less nested level).

```
> f(3);

f:
    1*    if g(x) < 25 then
           ...
          fi;
```

Execute the whole if-statement, including the call to the procedure g.

```
DBG> next
```
"less than five"

```
9
f:
    4    x^3

DBG> quit
```

The step debugger command also executes the current statement. However, execution stops before the next statement executed regardless of whether it is at the current nesting level, deeper, or shallower. In other words, the step command steps into nested statements and procedure calls.

```
> f(3);

f:
    1*    if g(x) < 25 then
           ...
```

```
        fi;
```

To evaluate the expression g(x)<25, Maple needs to call the procedure g, so the next statement displayed is in g.

```
DBG> step

g:
  1     2*x;
```

Inside g there is only one level, so the next and step debugger commands are equivalent.

```
DBG> next

6
g:
  2     x^2

DBG> step

f:
  2         print("less than five");
```

Having returned from g Maple is now inside the if statement.

```
DBG> quit
```

The into debugger command is functionally halfway between next and step. Execution stops at the next statement within the current procedure regardless of whether it is at the current nesting level, or within the body of a control structure (an if statement or a loop). In other words, the into command steps into nested statements but *not* into procedure calls.

```
> f(3);

f:
  1*     if g(x) < 25 then
             . . .
         fi;
```

Step into the if statement directly.

```
DBG> into

f:
  2         print("less than five");

DBG> quit
```

The outfrom debugger command causes execution within the current nesting level to complete. Execution stops again once it reaches a shallower level, that is, if a loop terminates, execution of a branch of an if statement terminates, or the current procedure call returns.

```
> f(3);

f:
    1*   if g(x) < 25 then
          ...
        fi;
```

Step into the if statement.

```
DBG> into

f:
    2        print("less than five");
```

Execute both statements in the body of the if statement.

```
DBG> outfrom
```

"less than five"

```
9
f:
    4    x^3

DBG> quit
```

The `return` debugger command causes execution of the currently active procedure call to complete. Execution stops once more at the first statement after the current procedure.

```
> f(3);

f:
    1*   if g(x) < 25 then
          ...
        fi;
```

Step into the procedure g.

```
DBG> step

g:
    1    2*x;
```

Return to f from g.

```
DBG> return

f:
    2        print("less than five");

DBG> quit
```

The help page ?debugger summarizes the capabilities of all the available debugger commands.

6.5 Restrictions

At the debugger prompt, the only Maple statements allowed are expressions, assignments, and quit (or done or stop). The debugger does not permit statements such as if, while, for, read, and save. However, you can use the 'if' operator to simulate an if statement, and the seq command to simulate a loop.

The debugger cannot set breakpoints in, or step into, kernel routines, such as diff and has; these are implemented in C and compiled into the Maple kernel. No debugging information about these routines is accessible to Maple since they deal with objects at a lower level than the debugger can deal with. You cannot use the debugger to debug the debugger, although it is written in the Maple language. This is not for reasons of secrecy (after all, you can use the debugger to debug Maple library functions), but rather to prevent an infinite recursion that would occur if you set a breakpoint in the debugger.

Finally, the debugger cannot determine with absolute certainty the statement at which execution stopped if a procedure contains two identical statements which also happen to be expressions. When this happens, you can still use the debugger and execution can continue. The debugger merely issues a warning that the displayed statement number may be incorrect. This problem occurs because Maple stores all identical expressions as a single occurrence of the expression, and the debugger has no way to determine at which invocation execution stopped.

Numerical Programming in Maple

Representing and manipulating expressions in symbolic mode; that is, in terms of variables, functions, and exact constants, is a powerful feature of the Maple system. However, practical scientific computation also demands *floating-point* calculations which represent quantities by approximate *numerical* values. Typically, numerical computations are used for one of three reasons.

First, not all problems have analytical or symbolic solutions. For example, of the many partial differential equations known, only a small subset have known closed-form solutions. But, you can usually find numerical solutions.

Second, the analytic answer that Maple returns to your problem may be very large or complex. You are not likely to do calculations by hand which involve rational numbers containing many digits or equations with hundreds of terms, but Maple does not mind such expressions. To understand big expressions, sometimes it helps to compute a floating-point approximation.

Third, you may not always need an exact answer. Computing an analytic answer of infinite precision is not necessary when your only interest is in an approximation. This situation typically arises in plotting. Calculating the points in the graph too accurately is wasteful because normal plotting devices are not capable of displaying ten digits of resolution.

While the rest of this book primarily shows Maple's powerful symbolic methods, the focus of this chapter is on how to perform floating-point calculations in Maple. You will quickly discover that Maple has some extraordinary capabilities in this regard. You have your choice of software floating-point calculations of arbitrary precision or hardware floating-point arithmetic. The former is unaffected, save for speed, by the machine you are using. The latter is determined by the architecture of your computer, but offers the advantage of exceptional speed.

7.1 The Basics of `evalf`

The `evalf` command is the primary tool in Maple for performing floating-point calculations. It causes Maple to evaluate in software floating-point mode. Maple's software floating-point arithmetic (see *Software Floats* on page 248) has an *n*-digit machine floating-point model as its basis, but allows computations at arbitrary precision. The environment variable `Digits`, which has an initial setting of 10, determines the default number of digits for calculations.

```
> evalf(Pi);
```

$$3.141592654$$

You may alter the number of digits either by changing the value of `Digits`, or by specifying the number as a second argument to `evalf`. Note that when you specify the number of digits as an argument to `evalf`, the default, `Digits`, remains unchanged.

```
> Digits := 20:
> evalf(Pi);
```

$$3.1415926535897932385$$

```
> evalf(Pi, 200);
```

$$3.141592653589793238462643383279502884 1\backslash$$
$$9716939937510582097494459230781640628 6\backslash$$
$$2089986280348253421170679821480865132 8\backslash$$
$$2306647093844609550582231725359408128 4\backslash$$
$$8111745028410270193852110555964462294 8\backslash$$
$$9549303820$$

```
> evalf(sqrt(2));
```

$$1.4142135623730950488$$

```
> Digits := 10:
```

The number of digits you specify is the number of *decimal* digits that Maple uses during calculations. Specifying a larger number of digits is likely to give you a more accurate answer, and the value of Digits has no theoretical limit. Depending on your computer, you may be able to perform calculations using half a million digits. Unlike most hardware implementations of floating-point arithmetic, Maple stores and performs software operations on floating-point numbers in base 10.

Results from evalf are not necessarily accurate to the setting of Digits. When you perform multiple operations, errors can accumulate—sometimes dramatically. You may be surprised that Maple does not automatically perform intermediate floating-point operations to greater accuracy to ensure that the result is as accurate as possible. Maple performs the floating-point calculations as *Software Floats* on page 248 describes. Maple deliberately limits its set of very accurate operations to those that you would encounter in normal arithmetic. In this way, you can use standard numerical methods to predict when results will be suspect. Much research has been done into the accuracy of floating-point calculations under the fixed number of digits model; texts and papers exist on many areas of the subject. Because Maple's model is so similar to the IEEE model, you can readily extend it to predict the accuracy of Maple calculations whenever extreme accuracy and reliability is of concern. Otherwise, if Maple tried to second-guess you, you would never know when Maple's model might fail and so you could *never* be certain of your results.

Sometimes a definite integral has no closed form solution in terms of standard mathematical functions. You can use evalf to obtain an answer via numerical integration.

```
> r := Int(exp(x^3), x=0..1);
```

$$r := \int_0^1 e^{(x^3)}\, dx$$

```
> value(r);
```

$$\int_0^1 e^{(x^3)}\, dx$$

```
> evalf(r);
```

$$1.341904418$$

In other cases, Maple can find an exact solution, but the form of the exact solution is almost incomprehensible. The function Beta below is one of the special functions that appear in the mathematical literature. Note that the Greek capital letter beta is almost indistinguishable from capital b.

```
> q := Int( x^99 * (1-x)^199 / Beta(100, 200), x=0..1/5 );
```

$$q := \int_0^{1/5} \frac{x^{99}\,(1-x)^{199}}{B(100,\,200)}\, dx$$

```
> value(q);
```

$$27852290545780521179255248650434305998 4 \backslash$$
$$03849800909690342170417622052715523897 \backslash$$
$$76190682816696442051841690247452471818 \backslash$$
$$79720294596176638677971757463413490644 \backslash$$
$$25727501861101435750157352018112989492 \backslash$$
$$972548449 \Big/ 21774128091037151646887 3\backslash$$
$$84971552115934384961767251671031013243 \backslash$$
$$12241148610308262514475552524051323083 \backslash$$
$$13238717840332750249360603782630341376 \backslash$$
$$82537367383346083183346165228661133571 \backslash$$
$$76260162148352832620593365691185012466 \backslash$$
$$14718189600663973041983050027165652595 \backslash$$
$$68426426994847133755683898925781250000 \backslash$$
$$0 \frac{1}{B(100,\ 200)}$$

```
> evalf(q);
```

$$.3546007367\ 10^{-7}$$

Note that the two examples above use the Int command rather than int for the integration. If you use int, Maple first tries to integrate your expression symbolically. Thus, when evaluating the commands below, Maple spends time finding a symbolic answer and then converts it to a floating-point approximation, rather than performing straight numerical integration.

```
> evalf( int(x^99 * (1-x)^199 / Beta(100, 200), x=0..1/5) );
```

$$.3546007367\ 10^{-7}$$

When you want Maple to perform numerical calculations, you should not use commands like int, limit, and sum that evaluate their arguments symbolically.

7.2   Hardware Floating-Point Numbers

Maple offers an alternative to software floating-point numbers: your computer's hardware floating-point arithmetic. Hardware floating-point calculations are typically much faster than software floating-point calculations. However, hardware floating-point arithmetic depends on your particular type of computer, and you cannot increase the precision.

The evalhf command evaluates an expression using hardware floating-point arithmetic.

```
> evalhf( 1/3 );
```

$$.333333333333333315$$

```
> evalhf( Pi );
```

$$3.14159265358979312$$

Your computer most likely does hardware floating-point arithmetic using a certain number of binary digits. The special construct, evalhf(Digits), approximates the corresponding number of decimal digits.

```
> d := evalhf(Digits);
```

$$d := 15.$$

Therefore, evalhf and evalf return similar results if evalf uses a setting of Digits that is close to evalhf(Digits). Maple usually shows you one or two digits more than the value of evalhf(Digits) specifies. When you perform hardware floating-point calculations, Maple must convert all the base-10 software floating-point numbers to base-2 hardware floating-point numbers, and then convert the result back to base 10. The extra decimal digits allow Maple to reproduce the binary number precisely if you use it again in a subsequent hardware floating-point calculation.

```
> expr := ln( 2 / Pi * ( exp(2)-1 ) );
```

$$expr := \ln\left(2\,\frac{e^2 - 1}{\pi}\right)$$

```
> evalhf( expr );
```

$$1.40300383684168617$$

```
> evalf( expr, round(d) );
```

$$1.40300383684169$$

The results that evalhf returns, even including for evalhf(Digits), are not affected by the value of Digits.

```
> Digits := 4658;
```

$$Digits := 4658$$

```
> evalhf( expr );
```

$$1.40300383684168617$$

```
> evalhf(Digits);
```

$$15.$$

```
> Digits := 10;
```

$$Digits := 10$$

You can use the `evalhf(Digits)` construct to tell whether hardware floating-point arithmetic provides sufficient precision in a particular application. If `Digits` is less than `evalhf(Digits)`, then you should take advantage of the faster hardware floating-point calculations; otherwise, you should use software floating-point arithmetic to perform the calculation, with sufficient digits. The `evaluate` procedure below takes an *unevaluated* parameter, expr. Without the `uneval` declaration, Maple would evaluate expr symbolically before invoking `evaluate`.

```
> evaluate := proc(expr::uneval)
>    if Digits < evalhf(Digits) then
>       evalhf(expr);
>    else
>       evalf(expr);
>    fi;
> end:
```

The `evalhf` command knows how to evaluate many of Maple's functions, but not all. For example, you cannot evaluate an integral using hardware floating-point arithmetic.

```
> evaluate( Int(exp(x^3), x=0..1) );
```

```
Error, (in evaluate)
unable to evaluate function 'Int' in evalhf
```

You can improve the `evaluate` procedure so that it traps such errors and tries to evaluate the expression using software floating-point numbers instead.

```
> evaluate := proc(expr::uneval)
>    local result;
>    if Digits < evalhf(Digits) then
>       result := traperror( evalhf(expr) );
>       if result = lasterror then
```

```
>               evalf(expr);
>           else
>               result;
>           fi;
>       else
>           evalf(expr);
>       fi;
> end:
> evaluate( Int(exp(x^3), x=0..1) );
```

$$1.341904418$$

The evaluate procedure provides a model of how to write procedures that take advantage of hardware floating-point arithmetic whenever possible.

Newton Iterations

You can use Newton's method to find numerical solutions to equations. As *Creating a Newton Iteration* on page 75 describes, if x_n is an approximate solution to the equation $f(x) = 0$, then x_{n+1}, given by the following formula, is typically a better approximation.

$$x_{n+1} = x_n - \frac{f(x_n)}{f'(x_n)}$$

This section illustrates how to take advantage of hardware floating-point arithmetic to calculate Newton iterations.

The iterate procedure below takes a function, f, its derivative, df, and an initial approximate solution, x0, as input to the equation $f(x) = 0$. iteration calculates at most N successive Newton iterations until the difference between the new approximation and the previous one is small. The iterate procedure prints out the sequence of approximations so you can follow the workings of the procedure.

```
> iterate := proc( f::procedure, df::procedure,
>                   x0::numeric, N::posint )
>   local xold, xnew;
>   xold := x0;
>   xnew := evalf( xold - f(xold)/df(xold) );
>   to  N-1 while abs(xnew-xold) > 10^(1-Digits) do
>       xold := xnew;
>       print(xold);
>       xnew := evalf( xold - f(xold)/df(xold) );
>   od;
>   xnew;
```

```
> end:
```

The procedure below calculates the derivative of f and passes all the necessary information to iterate.

```
> Newton := proc( f::procedure, x0::numeric, N::posint )
>     local df;
>     df := D(f);
>     print(x0);
>     iterate(f, df, x0, N);
> end:
```

Now you can use Newton to solve the equation $x^2 - 2 = 0$.

```
> f := x -> x^2 - 2;
```

$$f := x \to x^2 - 2$$

```
> Newton(f, 1.5, 15);
```

$$1.5$$

$$1.416666667$$

$$1.414215686$$

$$1.414213562$$

$$1.414213562$$

The version of Newton below uses hardware floating-point arithmetic if possible. Since iterate only tries to find a solution to an accuracy of 10^(1-Digits), Newton uses evalf to round the result of the hardware floating-point computation to an appropriate number of digits.

```
> Newton := proc( f::procedure, x0::numeric, N::posint )
>     local df, result;
>     df := D(f);
>     print(x0);
>     if Digits < evalhf(Digits) then
>         result := traperror(evalhf(iterate(f, df, x0, N)));
>         if result=lasterror then
>             iterate(f, df, x0, N);
>         else
>             evalf(result);
>         fi;
>     else
>         iterate(f, df, x0, N);
>     fi;
> end:
```

Below, Newton uses hardware floating-point arithmetic for the iterations and rounds the result to software precision. You can tell which numbers are hardware floating-point numbers because they have more digits than the software floating-point numbers, given the present setting of Digits.

```
> Newton(f, 1.5, 15);
```

$$1.5$$

$$1.41666666666666674$$

$$1.41421568627450989$$

$$1.41421356237468987$$

$$1.41421356237309515$$

$$1.414213562$$

You may find it surprising that Newton must use software floating-point arithmetic to find a root of the Bessel function below.

```
> F := z -> BesselJ(1, z);
```

$$F := z \rightarrow \text{BesselJ}(1, z)$$

```
> Newton(F, 4, 15);
```

$$4$$

$$3.826493523$$

$$3.831702467$$

$$3.831705970$$

$$3.831705970$$

The reason is that evalf does not know about BesselJ and the symbolic code for BesselJ uses the type command, which evalhf does not allow.

```
> evalhf( BesselJ(1, 4) );
```

```
Error, unsupported type in evalhf
```

Using traperror, as in the Newton procedure above, allows your procedure to work even when evalhf fails.

You may wonder why the Newton procedure above prints out so many digits when it is trying to find a ten-digit approximation. The reason is that the print command is located inside the iterate procedure which is inside a call to evalhf, where all numbers are hardware floating-point numbers, and print as such.

Computing with Arrays of Numbers

Use the evalhf command for calculations with numbers. The only structured Maple object allowed in a call to evalhf is arrays of numbers. If an array has undefined entries, `evalhf` initializes them to zero. The procedure below calculates the polynomial $2 + 5x + 4x^2$.

```
> p := proc(x)
>    local a, i;
>    a := array(0..2);
>    a[0] := 2;
>    a[1] := 5;
>    a[2] := 4;
>    sum( a[i]*x^i, i=0..2 );
> end:
> p(x);
```

$$2 + 5x + 4x^2$$

If you intend to enclose p in a call to `evalhf`, you cannot define the local array a using `array(1..3, [2,5,4])`, since lists are not allowed inside `evalhf`. You can, however, enclose p in a call to `evalhf` if the parameter x is a number.

```
> evalhf(p(5.6));
```

$$155.439999999999998$$

You can also pass an array of numbers as a parameter inside a call to `evalhf`. The procedure below calculates the determinant of a 2×2 matrix. The $(2, 2)$ entry in the array a below is undefined.

```
> det := proc(a::array(2))
>    a[1,1] * a[2,2] - a[1,2] * a[2,1];
> end:
> a := array( [[2/3, 3/4], [4/9]] );
```

$$a := \begin{bmatrix} \frac{2}{3} & \frac{3}{4} \\ [2ex]\frac{4}{9} & a_{2,2} \end{bmatrix}$$

```
> det(a);
```

$$\frac{2}{3} a_{2,2} - \frac{1}{3}$$

If you call det from inside a call to `evalhf`, Maple uses the value 0 for the undefined entry, `a[2,2]`.

```
> evalhf( det(a) );
```

$$-.333333333333333315$$

`evalhf` passes arrays by value, so the $(2, 2)$ entry of a is still undefined.

```
> a[2,2];
```

$$a_{2,2}$$

If you want `evalhf` to modify an array that you pass as a parameter to a procedure, you must enclose the name of the array in a `var` construct. The var construct is special to `evalhf` and is necessary only if you want `evalhf` to modify an array of numbers that is accessible at the session level.

```
> evalhf( det( var(a) ) );
```

$$-.333333333333333315$$

Now a is an array of floating-point numbers.

```
> eval(a);
```

$$[.666666666666666630 , .750000000000000000]$$

$$[.444444444444444420 , 0]$$

The `evalhf` command always returns a single floating-point number, but the `var` construct allows you to calculate a whole array of numbers with one call to `evalhf`. *Generating Grids of Points* on page 309 illustrates the use of `var` to calculate a grid of function values that you can use for plotting.

You can also create arrays of hardware floating-point values directly with the `hfarray` command. Proper use of this command can save significant amounts of time, especially in plotting routines, which rely heavily on arrays of floating-point values. See the help page for `hfarray` for more details and examples.

7.3 Floating-Point Models in Maple

You can represent a real number in floating-point notation. You simply write a series of digits and insert a point (such as a period or comma) to mark the division between digits which represent whole and fractional numbers. When you attempt this on paper you will likely choose to write the numbers base 10, but most computer hardware uses base 2. For example, if you were to write the number 34.5, you mean the following:

$$3 \times 10^1 + 4 \times 10^0 + 5 \times 10^{-1}.$$

Sometimes you may wish to write numbers in *scientific notation*, especially if they are very large or very small. This means that you write the number as a number greater than or equal to 1 and less than 10, so that the first digit is not zero, multiplied by some power of ten. Thus, you would write

34.5 as

$$3.45 \times 10^1.$$

The term *floating-point* arises from the fact that to represent a real number you must *float the point* to place it just to the right of the first non-zero digit.

Of course, to represent all numbers you must use an infinite number of digits, because many numbers, such as π, require an infinite number of digits. In fact, representing real numbers in floating-point notation is difficult for two reasons: the real numbers are a *continuum* (if x is a number between two real numbers, then x is also real) and because, like the integers, the reals are *unbounded* (no number is the largest).

A *floating-point model* has emerged in recent years defining how to design hardware implementations. This standard covers not only base-2 arithmetic, since most computer chips store numbers in binary format, but also covers operations using other bases. You can represent each non-zero number in base β as

$$(-1)^{sign} \times (d_0.d_1d_2 \ldots d_{Digits-1}) \times \beta^e.$$

A special representation exists for zero. The limit on the magnitude of the exponent, e, depends on how much space you reserve to store it. In general, it occurs between some lower limit, L, and some upper limit, U, inclusive. The digits, $d_0.d_1d_2 \ldots d_{Digits-1}$ are called the the *significand*.

Software Floats

The base for Maple software floats is $\beta = 10$. The choice of a decimal base, rather than a binary base which is common in hardware implementations, allows the arithmetic to model closely the arithmetic you do by hand. Furthermore, no conversion is necessary when printing these numbers.

A significant feature of software floats is that you may specify the precision, *Digits*, to any positive integer simply by assigning to the Maple environment variable Digits. The maximum value is implementation dependent, but is generally larger than you would encounter in any practical computation—typically greater than 500 000 decimal digits.

The largest single precision integer in the underlying computer system is the limit of the exponent of software floats. On a typical 32-bit computer, this number is $2^{31} - 1$ and so the limits are $L = -(2^{31} - 1)$ and $U = 2^{31} - 1$.

The main difference between the Maple format and the standard floating-point model is that the sequence of digits representing the significand has an implied decimal point following the *last* digit, rather than immediately following the first digit. Thus 34.5 is represented by Maple as Float(345, -1).

Hardware Floats

You may also use the floating-point system built into your computer by means of the evalhf command. In this case, Maple uses the double-precision floating-point system of your computer. Naturally, the specifics depend upon your computer. A typical 32-bit computer uses base $\beta = 2$ and uses two words, that is 64 bits, to store a double-precision float.

The IEEE Floating-Point Standard for binary double-precision floating-point arithmetic on a 32-bit computer stipulates that Maple reserves one bit to store the sign and eleven bits to store the exponent. The standard further specifies that the bounds on the exponent are $L = -1022$ and $U = 1023$. The number of bits remaining (out of 64 bits) for representing the significand is 52. This seems to imply that the precision is *binaryDigits* = 52. However, the leading digit cannot be zero and so in a binary system must always be one. Thus, the standard specifies that *binaryDigits* = 53 and that the leading bit, d_0, is an *implicit bit* with value 1. The number zero is represented by an exceptional value encoded in the exponent field.

The 53-bit precision of the binary representation corresponds to approximately sixteen decimal digits. The largest positive number, converted to decimal notation, is just less than

$$2 \times 2^{1023} \approx 1.8 \times 10^{308}$$

and the smallest positive number is

$$2^{-1022} \approx 2.2 \times 10^{-308}.$$

Roundoff Error

When you perform floating-point arithmetic, whether using software or hardware floats, you are using *approximate* numbers rather than precise real numbers or expressions. Maple can work with exact (symbolic) expressions. The difference between an exact real number and its floating-point approximation is called the *roundoff error*. For example, suppose you request a floating-point representation of π.

```
> pie := evalf(Pi);
```

$$pie := 3.141592654$$

Maple rounds the precise value π to ten significant digits because `Digits` is set to its default value of 10. You can approximate the roundoff error above by temporarily increasing the value of `Digits` to 15.

```
> evalf(Pi - pie, 15);
```

$$-.41021\ 10^{-9}$$

Roundoff errors arise not only from the representation of input data, but also as a result of performing arithmetic operations. Each time you perform an arithmetic operation on two floating-point numbers, the infinitely-precise result usually will not be representable in the floating-point number system and therefore the computed result will also have an associated roundoff error.

For example, suppose you multiply two ten-digit numbers. The result can easily have nineteen or twenty digits, but Maple can only store the first ten digits in a ten-digit floating-point system.

```
> 1234567890 * 1937128552;
```

$$2391516709101395280$$

```
> evalf(1234567890) * evalf(1937128552);
```

$$.2391516709\ 10^{19}$$

Maple adheres to the floating-point standard, which states that whenever you apply one of the four basic arithmetic operations (addition, subtraction, multiplication, or division) to two floating-point numbers, then the result is the correctly rounded representation of the infinitely-precise result, unless overflow or underflow occurs. Of course, Maple may need to compute an extra digit or two behind the scenes to ensure that the answer is correct.

Even so, sometimes a surprising amount of error can accumulate, particularly when subtracting two numbers which are of similar magnitude. In the calculation below, the accurate sum of x, y, and z is $y = 3.141592654$.

```
> x := evalf(987654321);
```

$$x := .987654321\ 10^{9}$$

```
> y := evalf(Pi);
```

$$y := 3.141592654$$

```
> z := -x;
```

$$z := -.987654321\ 10^{9}$$

```
> x + y + z;
```

$$3.1$$

Catastrophic cancellation is the name of this phenomenon. During the subtraction the eight leading digits cancel out, leaving only two significant digits in the result.

One advantage of Maple's software floats, in contrast to fixed-precision floating-point numbers systems, is that the user can increase the precision to alleviate some of the consequences of roundoff errors. For example, increasing Digits to 20 dramatically improves the result.

```
> Digits := 20;
```

$$Digits := 20$$

```
> x + y + z;
```

$$3.141592654$$

You should employ standard numerical analysis techniques to avoid large errors accumulating in your calculations. Often, reordering the operations leads to a more accurate final result. For example, when computing a sum, add the numbers with the smallest magnitude first.

7.4 Extending the `evalf` Command

The `evalf` command knows how to evaluate many functions and constants, such as `sin` and `Pi`. You can also define your own functions or constants, and extend `evalf` by adding information about how to compute such functions or constants.

Defining Your Own Constants

You may define a new constant and write procedures that manipulate this constant symbolically. You should then write a procedure that can calculate a floating-point approximation of your constant to any number of digits. If you assign the procedure a name of the form `'evalf/constant/name'`, then Maple invokes the procedure when you use `evalf` to evaluate an expression containing your constant, *name*.

Suppose you want the name MyConst to represent the following infinite series:

$$MyConst = \sum_{i=1}^{\infty} \frac{(-1)^i \pi^i}{2^i i!}.$$

You can calculate approximations using the above series in many ways; the procedure below is one implementation. Note that if a_i is the ith term in the sum, then $a_{i+1} = -a_i(\pi/2)/i$ gives the next term. You can calculate an approximation to the series by adding terms until Maple's model for software floating-point numbers cannot distinguish the new partial sum from the previous one. Using numerical analysis, you can prove that this algorithm calculates Digits accurate digits of MyConst, if you use two extra digits inside the algorithm. Therefore, the procedure below increments Digits by two and uses evalf to round the result to the proper number of digits before returning. The procedure does not have to reset the value of Digits because Digits is an environment variable.

```
> 'evalf/constant/MyConst' := proc()
>     local i, term, halfpi, s, old_s;
>     Digits := Digits + 2;
>     halfpi := evalf(Pi/2);
>     old_s := 1;
>     term := 1.0;
>     s := 0;
>     for i from 1 while s <> old_s do
>         term := -term * halfpi / i;
>         old_s := s;
>         s := s + term;
>     od;
>     evalf(s, Digits-2);
> end:
```

When you invoke evalf on an expression containing MyConst, Maple invokes 'evalf/constants/MyConst' to calculate an approximate value.

```
> evalf(MyConst);
```

$$-.7921204236$$

```
> evalf(MyConst, 40);
```

$$-.7921204236492380914530443801650212299661$$

You can express the particular constant, MyConst, in closed form and, in this case, you can use the closed-form formula to calculate approximations to MyConst more efficiently.

```
> Sum( (-1)^i * Pi^i / 2^i / i!, i=1..infinity );
```

$$\sum_{i=1}^{\infty} \frac{(-1)^i \, \pi^i}{2^i \, i!}$$

```
> value(%);
```

$$e^{(-1/2\,\pi)} \left(1 - e^{(1/2\,\pi)}\right)$$

```
> expand(%);
```

$$\frac{1}{\sqrt{e^\pi}} - 1$$

```
> evalf(%);
```

$$-.7921204237$$

Defining Your Own Functions

If you define your own functions, you may want to write your own procedure for calculating numerical approximations to the function values. When you invoke `evalf` on an expression containing an unevaluated call to a function F, then Maple calls the procedure `'evalf/F'` if such a procedure exists.

Suppose you want to study the function $x \mapsto (x - \sin(x))/x^3$.

```
> MyFcn := x -> (x - sin(x)) / x^3;
```

$$MyFcn := x \to \frac{x - \sin(x)}{x^3}$$

This function is not defined at $x = 0$, but you can extend it as a continuous function by placing the limiting value in MyFcn's remember table.

```
> MyFcn(0) := limit( MyFcn(x), x=0 );
```

$$MyFcn(0) := \frac{1}{6}$$

For small values of x, $\sin(x)$ is almost equal to x, so the subtraction $x - \sin(x)$ in the definition of MyFcn can lead to inaccuracies due to catastrophic cancellation. When you evaluate v below to ten digits, only the first two are correct.

```
> v := 'MyFcn'( 0.000195 );
```

$$v := MyFcn(.000195)$$

```
> evalf(v);
```

$$.1618368482$$

```
> evalf(v, 2*Digits);
```

$$.16666666634973617222$$

If you depend on accurate numerical approximations of MyFcn, you must write your own procedure to provide them. You could write such a procedure by exploiting the series expansion of MyFcn.

```
> series( MyFcn(x), x=0, 11 );
```

$$\frac{1}{6} - \frac{1}{120}x^2 + \frac{1}{5040}x^4 - \frac{1}{362880}x^6 + O(x^8)$$

The general term in the series is

$$a_i = (-1)^i \frac{x^{2i}}{(2i+3)!}, \qquad i \geq 0.$$

Note that $a_i = -a_{i-1}x^2/((2i+2)(2i+3))$. For small values of x, you can then calculate an approximation to MyFcn(x) by adding terms until Maple's model for software floating-point numbers cannot distinguish the new partial sum from the previous one. For largers values of x, catastrophic cancellation is not a problem, so you can use evalf to evaluate the expression. Using numerical analysis, you can prove that this algorithm calculates Digits accurate digits of the function value if you use three extra digits inside the algorithm. Therefore, the procedure below increments Digits by three and uses evalf to round the result to the proper number of digits before returning.

```
> 'evalf/MyFcn' := proc(xx::algebraic)
>     local x, term, s, old_s, xsqr, i;
>     x := evalf(xx);
>     Digits := Digits+3;
>     if type(x, numeric) and abs(x)<0.1 then
>         xsqr := x^2;
>         term := evalf(1/6);
>         s := term;
>         old_s := 0;
>         for i from 1 while s <> old_s do
>             term := -term * xsqr / ((2*i+2)*(2*i+3));
>             old_s := s;
>             s := s + term;
>         od;
>     else
>         s := evalf( (x-sin(x))/x^3 );
>     fi;
>     evalf(s, Digits-3);
> end:
```

When you invoke evalf on an expression containing an unevaluated call to MyFcn, Maple invokes 'evalf/MyFcn'.

```
> evalf( 'MyFcn'(0.000195) );
```

$$.1666666663$$

You should now recode the symbolic version of `MyFcn` so that it takes advantage of `'evalf/MyFcn'` if the argument is a floating-point number.

```
> MyFcn := proc(x::algebraic)
>    if type(x, float) then
>       evalf('MyFcn'(x));
>    else
>       (x - sin(x)) / x^3;
>    fi;
> end:
```

The `evalf` command automatically looks for `'evalf/MyFcn'` when used in the `evalf('MyFcn')` syntax.

```
> MyFcn(0) := limit( MyFcn(x), x=0 );
```

$$\text{MyFcn}(0) := \frac{1}{6}$$

Now you can properly evaluate `MyFcn` with numeric as well as symbolic arguments.

```
> MyFcn(x);
```

$$\frac{x - \sin(x)}{x^3}$$

```
> MyFcn(0.099999999);
```

$$.1665833532$$

```
> MyFcn(0.1);
```

$$.1665833532$$

Extending Maple on page 89 describes how to extend many other Maple commands.

7.5 Using the Matlab Package

Another way to accomplish numerical computations in Maple is to use the `Matlab` package which provides a way to access several of Matlab's built-in functions (assuming you have a copy of Matlab properly installed on your computer). The mathematical functions provided are:

- `chol`: Cholesky factorization
- `det`: determinant
- `eig`: eigenvalues and eigenvectors
- `fft`: discrete Fourier transforms

- `inv`: matrix inverse
- `lu`: LU decomposition
- `ode45`: solve ordinary differential equation
- `qr`: QR orthogonal-triangular decomposition
- `transpose`: matrix transposition

There are also a handful of support/utility commands provided.

Matlab converts all Maple structures to `hfarrays` — arrays of hardware floating-point values — before it performs any computations. The results you get will usually be in terms of hfarrays, not standard Maple matrices. The `convert( ,array)` command has been extended to handle any conversions between the two.

For more information on all these commands and the `Matlab` package in general, please refer to the on-line help for `Matlab`. To learn how to start the matlab application from within your Maple session, see the on-line help for `Matlab[openlink]`.

7.6   Conclusion

The various techniques described in this chapter afford an important extension to Maple's programming language and its ability to perform symbolic manipulations. With numerical techniques at your disposal, you can do the following: solve equations which are otherwise unsolvable, investigate the properties of complicated solutions, and quickly obtain numerical estimates.

Symbolic calculations give precise representations, but in some cases can be expensive to compute even with such a powerful tool as Maple. At the other extreme, hardware floating-point arithmetic allows you fast computation directly from Maple. This involves, however, limited accuracy. Software floating-point offers a balance. As well as sometimes being much faster than symbolic calculations, you also have the option to control the precision of your calculations, thus exerting control over errors.

Software floating-point calculations and representations mimic the IEEE standard representation closely, except for the great advantage of arbitrary precision. Because of the similarity with this popular standard, you can readily apply the knowledge of accumulation of error and numerical analysis principles that numerous texts and papers contain. When you need to know that your calculations are precise, this wealth of information at your disposal should provide you with confidence in your results.

Programming with Maple Graphics

8

Maple has a wide range of commands for generating both two- and three-dimensional plots. For mathematical expressions, you can use library procedures, such as `plot` and `plot3d`, or one of the many specialized graphics routines found in the `plots` and `plottools` packages, the `DEtools` package (for working with differential equations), and the `stats` package (for statistical data). The input to these commands is typically one or more Maple formulae, operators, or functions, along with information about domains and possibly ranges. In all cases, the graphic commands allow for the setting of options, specifying such attributes as coloring, shading, or axes style.

The purpose of this chapter is to reveal the structure of the procedures that Maple uses to generate graphical output, and allow you to generate your own graphics procedures. This chapter includes basic information about argument conventions, setting defaults, and processing of plotting options. A major part of the material describes the data structures that Maple uses for plotting along with various techniques to build such data structures in order to produce graphics in Maple. In addition, you will see how some of the existing functions in the `plots` and `plottools` packages produce specific plotting data structures.

8.1 Basic Plot Functions

This section illustrates some of the basic workings of the graphics procedures in Maple, as well as some of the properties that are common to all Maple plotting commands. Also, it discusses plotting Maple operators or functions versus formula expressions and the setting of optional information.

Several of Maple's graphics procedures take mathematical expressions as their input. Examples of such commands include `plot`, `plot3d`, `animate`, `animate3d`, and `complexplot`. All these commands allow the input to be in one of two forms: formulae or functions. The former consists of expressions such as $x^2y - y^3 + 1$ or $3\sin(x)\sin(y) + x$, both formulae in the variables x and y. If p and q are functions with two arguments, then $p+q$ is an example of a function expression. The graphics procedures use the way you specify the domain information to determine if the input is a function expression or a formula in a specified set of variables. For example, the command below generates a three-dimensional plot of the surface which $\sin(x)\sin(y)$ defines. This formula is in terms of x and y.

```
> plot3d( sin(x) * sin(y), x=0..4*Pi, y=-2*Pi..2*Pi );
```

If instead, you define two functions, each with two arguments,

```
> p := (x, y) -> sin(x):   q := (x, y) -> sin(y):
```

then you can plot the surface $p * q$ determines in the following manner.

```
> plot3d( p * q, 0..4*Pi, -2*Pi..2*Pi );
```

Both cases produce the same three-dimensional plot. In the first example, you supply the information that the input is an expression in x and y by giving the second and third arguments in the form x = *range* and y = *range*, while in the second example, there are no variable names.

Working with formula expressions is simple, but in many cases, functions provide a better mechanism for constructing mathematical functions. The following constructs a mathematical function which, for a given input, computes the required number of iterations (to a maximum of 10) for the sequence $z_{n+1} = z_n^2 + c$ to exit the disk of radius 2 for various complex starting points $c = x + iy$.

```
> mandelbrotSet := proc(x, y)
>     local z, m;
>     z := evalf( x + y*I );
>     m := 0;
>     to 10 while abs(z) < 2 do
>         z := z^2 + (x+y*I);
>         m := m + 1;
>     od:
>     m;
> end:
```

You now have a convenient method for computing a three-dimensional Mandelbrot set on a 50×50 grid.

```
> plot3d( mandelbrotSet, -3/2..3/2, -3/2..3/2, grid=[50,50] );
```

Creating a Maple graphic at the command level displays it on the plotting device (that is, your terminal). In many cases, you can then interactively alter the graph using the tools available with these plotting devices. Examples of such alterations include changing the drawing style, the axes style, and the view point. You can include this information by using optional arguments to plot3d.

```
> plot3d( sin(x)*sin(y), x=-2*Pi..2*Pi, y=-2*Pi..2*Pi,
>          style=patchnogrid, axes=frame );
```

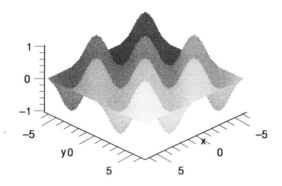

```
> plot3d( mandelbrotSet, -1.5..1.5, -1.5..1.5, grid=[50,50],
>          style=wireframe, orientation=[143,31] );
```

Every plotting procedure allows for optional arguments. You give the optional information in the form *name=option*. Some of these options affect the amount of information concerning the function that you give to the plotting procedures. The grid option that the Mandelbrot set example uses is an example. You can use other options for specifying visual information once you have determined the graphical points. The type of axes, shading, surface style, line styles, and coloring are but a few of the options available in this category. Obtain information about all the allowable options for the two-dimensional and three-dimensional cases using the help pages ?plot,options and ?plot3d,options.

Any graphics routine you create should allow users a similar set of options. When writing programs that call existing Maple graphics routines, simply pass the potential optional arguments directly to these routines.

8.2 Programming with Plotting Library Functions

This section gives examples of programming with the graphics procedures in Maple.

Plotting a Loop

Consider the first problem of plotting a loop from a list of data.

```
> L1 := [ [5,29], [11,23], [11,36], [9,35] ];
```

$$L1 := [[5, 29], [11, 23], [11, 36], [9, 35]]$$

The plot command draws lines between the listed points.

```
> plot( L1 );
```

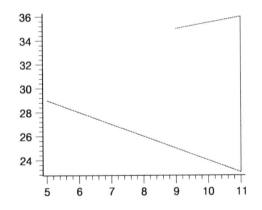

You may want to write a procedure that also draws a line from the last to the first point. All you need to do is append the first point in L1 to the end of L1.

```
> L2 := [ op(L1), L1[1] ];
```

$$L2 := [[5, 29], [11, 23], [11, 36], [9, 35], [5, 29]]$$

```
> plot( L2 );
```

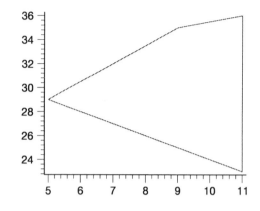

The procedure `loopplot` automates this technique.

```
> loopplot := proc( L )
>    plot( [ op(L), L[1] ] );
> end;
```

$$loopplot := \mathbf{proc}(L) \, plot([op \, (L), \, L_1]) \; \mathbf{end}$$

This procedure has a number of shortcomings. You should always verify the input, L, to `loopplot` to be a list of points, where a point is a list of two constants. That is, L should be of type `list([constant, constant])`. The `loopplot` command should also allow a number of plotting options. All `loopplot` has to do is pass on the options to `plot`. Inside a procedure, `args` is the sequence of arguments in the call to the procedure, and `nargs` is the number of arguments. Thus `args[2..nargs]` is the sequence of options passed to `loopplot`. The `loopplot` procedure should pass all but its first argument, L, directly to `plots`.

```
> loopplot := proc( L::list( [constant, constant] ) )
>    plot( [ op(L), L[1] ], args[2..nargs] );
> end:
```

The above version of `loopplot` gives an informative error message if you try to use it with improper arguments, and it also allows plotting options.

```
> loopplot( [[1, 2], [a, b]] );

Error, loopplot expects its 1st argument, L,
to be of type list([constant, constant]),
```

```
but received [[1, 2], [a, b]]

> loopplot( L1, linestyle=3 );
```

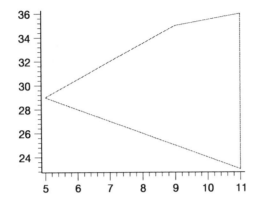

You may want to improve `loopplot` so that it can handle the empty list as input.

A Ribbon Plot Procedure

This section ends with the creation of a `ribbonplot` procedure, a three-dimensional plot of a list of two-dimensional formulae or functions. This is a first attempt at writing `ribbonplot`.

```
> ribbonplot := proc( Flist, r1 )
>    local i, m, p, y;
>    m  := nops(Flist);
>  # Create m translated plots.
>    p := seq( plot3d( Flist[i], r1, y=(i-1)..i ), i=1..m );
>    plots[display]( p );
> end:
```

The `ribbonplot` procedure uses the the `display` procedure from the `plots` package to display the plots. This procedure is called explicitly using its full name so that `ribbonplot` will work even when the short names for the functions in the `plots` package have not been loaded.

Now you can try out the procedure.

```
> ribbonplot( [cos(x), cos(2*x), sin(x), sin(2*x)],
>    x=-Pi..Pi );
```

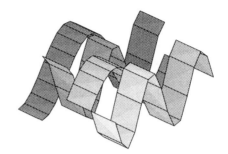

The above ribbonplot procedure uses too many grid-points in the
y-direction, two are sufficient. Thus, you need a grid=[*numpoints*, 2]
option to the plot3d command in ribbonplot. Here *numpoints* is the
number of points that ribbonplot should use in the x-direction; which
you should set with an option to ribbonplot. The hasoption command
helps you handle options. In the ribbonplot procedure below, hasoption
returns false if numpoints is not among the options listed in opts. If
opts contains a numpoints option, then hasoption assigns the value of
the numpoints option to n, and returns the remaining options in the fourth
argument (in this case modifying the value of the list opts).

```
> ribbonplot := proc( Flist, r1::name=range )
>    local i, m, p, y, n, opts;
>    opts := [ args[3..nargs] ];
>    if not hasoption( opts, 'numpoints', 'n', 'opts' )
>    then n := 25 # default numpoints
>    fi;
>
>    m := nops( Flist );
>    # op(opts) is any additional options
>    p := seq( plot3d( Flist[i], r1, y=(i-1)..i,
>                      grid=[n, 2], op(opts) ),
>              i=1..m );
>    plots[display]( p );
> end:
```

Now ribbonplot uses the number of grid points you ask it to.

```
> ribbonplot( [cos(x), cos(2*x), sin(x), sin(2*x)],
>                x=-Pi..Pi, numpoints=16 );
```

The input to ribbonplot above must be a list of expressions. You should extend ribbonplot so that it also accepts a list of functions. One difficulty with this extension is that you need to create two-dimensional functions from one-dimensional functions, something that was not a problem in the initial examples of ribbonplot. For this you can create an auxiliary procedure, extend, which makes use of the unapply command.

```
> extend := proc(f)
>    local x,y;
>    unapply(f(x), x, y);
> end:
```

For example, the extend procedure converts the $R \to R$ function $x \mapsto \cos(2x)$ to a $R^2 \to R$ function.

```
> p := x -> cos(2*x):
> q := extend(p);
```

$$q := (x, y) \to \cos(2\,x)$$

The following gives the new ribbonplot code.

```
> ribbonplot := proc( Flist, r1::{range, name=range} )
>    local i, m, p, n, opts, newFlist;
>    opts := [ args[3..nargs] ];
>    if type(r1, range) then
>     #  Functional input.
>       if not hasoption( opts, 'numpoints', 'n', 'opts' )
>       then n := 25 # default numpoints
>       fi;
>       m := nops( Flist );
>     #  change plot3d for functional input
```

```
>       p := seq( plot3d( extend( Flist[i] ), r1, (i-1)..i,
>                        grid=[n,2], op(opts) ),
>              i=1..m );
>       plots[display]( p );
>    else
>     #  Expressions. Convert each to a function of lhs(r1).
>       newFlist := map( unapply, Flist, lhs(r1) );
>     #  Use lhs(r1) as the default x-axis label.
>       opts := [ 'labels'=[lhs(r1), "", "" ],
>               args[3..nargs] ];
>       ribbonplot( newFlist, rhs(r1), op(opts) )
>    fi
> end:
```

Here is a ribbon plot of three functions.

```
> ribbonplot( [cos, sin, cos + sin], -Pi..Pi );
```

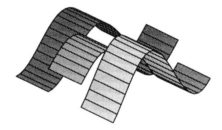

8.3 Maple's Plotting Data Structures

Maple generates plots by sending the user interface an unevaluated PLOT or PLOT3D function call. The information included inside these functions determines the objects they will graph. Every command in the plots package creates such a function. View this flow of information in the following manner. A Maple command produces a PLOT structure and passes it to the user interface. In the user interface, Maple constructs primitive graphic objects based on the PLOT structure. It then passes these objects to the chosen device driver for display. This process is shown schematically in figure 8.1.

You can assign the plotting data structures to variables, transform them into other structures, save them, or even print them out.

You can see examples of a plot structure in either two- or three-dimensions by line printing such a structure.

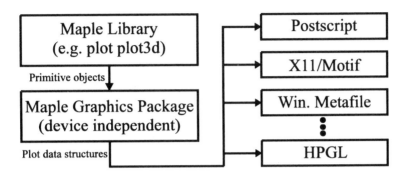

FIGURE 8.1 HOW PLOTS ARE DISPLAYED

```
> lprint( plot(2*x+3, x=0..5, numpoints=3, adaptive=false) );

PLOT(CURVES([[0, 3.], [2.61565849999999989, 8.23131700\
000000066], [5., 13.]],COLOUR(RGB,1.0,0,0)),AXESLABELS
("x",''),VIEW(0 .. 5.,DEFAULT))
```

Here, plot generates a PLOT data structure that includes the information
for a single curve defined by three points, with the curve colored with the
red-green-blue (RGB) values (1.0, 0, 0), which corresponds to red. The plot
has a horizontal axis running from 0 to 5. Maple, by default, determines
the scale along the vertical axes using the information that you provide in
the vertical components of the curve. The numpoints = 3 and adaptive
= false settings ensure that the curve consists of only three points.

The second example is the graph of $z = xy$ over a 3×4 grid. The
PLOT3D structure contains a grid of z values over the rectangular region
$[0, 1] \times [0, 2]$.

```
> lprint( plot3d(x*y, x=0..1, y=0..2, grid=[3,4]) );

PLOT3D(GRID(0 .. 1.,0 .. 2.,hfarray(1..3,1..4,[[0,0,0,
0],[0,0.333333333333333,0.666666666666667,1],[0,
0.666666666666667,1.33333333333333,2]])),STYLE(PATCH),
```

```
AXESLABELS(x,y,''))
```

The structure includes labels x and y for the plane but no label for the z-axis.

The third example is again the graph of $z = xy$ but this time in cylindrical coordinates. The PLOT3D structure now contains a mesh of points that make up the surface, along with the information that the plotting device should display the surface in a point style.

```
> lprint( plot3d( x*y, x=0..1, y=0..2, grid=[3,2],
>                  coords=cylindrical, style=point ) );
```

```
PLOT3D(MESH(hfarray(1..3,1..2,1..3,[[[0,0,0],[0,0,2]],
[[0,0,0],[0.877582561890373,0.479425538604203,2]],[[0,
0,0],[1.08060461173628,1.68294196961579,2]]])),STYLE(
POINT))
```

Since the plot is not in cartesian coordinates there are no default labels, whence the PLOT3D structure does not contain any AXESLABELS.

The PLOT Data Structure

You can construct and manipulate a plotting data structure directly to create two- and three-dimensional plots. All you need is a correct arrangement of the geometric information inside a PLOT or PLOT3D function. The information inside this function determines the objects that the plotting device displays. Here Maple evaluates the expression

```
> PLOT( CURVES( [ [0,0], [2,1] ] ) );
```

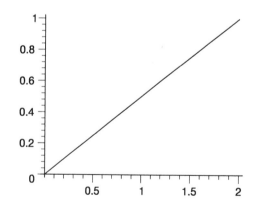

and passes it to the Maple interface which determines that this is a plot data structure. The Maple interface then dismantles the contents and passes the information to a plot driver which then determines the graphical information that it will render onto the plotting device. In the latest example, the result is a single line from the origin to the point (2, 1). The CURVES data structure consists of one or more lists of points each generating a curve, along with some optional arguments (for example, line style or line thickness information). Thus, the expression

```
> n := 200:
> points := [ seq( [2*cos(i*Pi/n), sin(i*Pi/n) ], i=0..n) ]:
> PLOT( CURVES( evalf(points) ) );
```

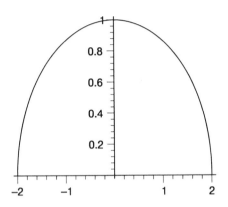

generates the plot of a sequence of $n + 1$ points in the plane. The points found inside the PLOT data structure must be numeric. If you omit the evalf statement, then non-numeric objects within the PLOT structure, such as $\sin(\pi/200)$, cause an error.

```
> PLOT( CURVES( points ) );
```

```
Error in iris-plot: Non-numeric vertex definition
```

```
> type( sin(Pi/n), numeric );
```

$$false$$

Hence, no plot is generated.

In general, the arguments inside a PLOT structure are all of the form

ObjectName(ObjectInformation , LocalInformation)

where *ObjectName* is a function name, for example one of CURVES, POLYGONS, POINTS, or TEXT; *ObjectInformation* contains the basic geometric point information that describes the particular object; and the optional *Local-Information* contains information about options that apply only to this particular object. *ObjectInformation* depends on the *ObjectName*. In the case where the *ObjectName* is CURVES or POINTS, the *ObjectInformation* consists of one or more lists of two-dimensional points. Each list supplies the set of points making up a single curve in the plane. Similarly, when *ObjectName* is POLYGONS, then the object information consists of one or more lists of points where each list describes the vertices of a single polygon in the plane. When *ObjectName* is TEXT, the object information consists of a point location along with a text string. The optional information is also in the form of an unevaluated function call. In the two-dimensional case, the options include AXESSTYLE, STYLE, LINESTYLE, THICKNESS, SYMBOL, FONT, AXESTICKS, AXESLABELS, VIEW, and SCALING.

You can also place some of these as *LocalInformation* inside a POINTS, CURVES, TEXT, or POLYGONS object; *LocalInformation* overrides the global option for the rendering of that object. The COLOR option allows for a further format when you place it on an object. In the case of an object having multiple subobjects (for example multiple points, lines, or polygons), you can supply one color value for each object.

Here is a simple way to generate a filled histogram of sixty-three values of the function $y = \sin(x)$ from 0 to 6.3. Maple colors each trapezoid individually by the HUE value corresponding to $y = |\cos(x)|$.

```
> p := i -> [ [(i-1)/10, 0], [(i-1)/10, sin((i-1)/10)],
>               [i/10, sin(i/10)], [i/10, 0] ]:
```

Now p(i) is the list of corners of the ith trapezoid. For example, p(2) contains the corners of the second trapezoid.

```
> p(2);
```

$$\left[\left[\frac{1}{10}, 0\right], \left[\frac{1}{10}, \sin\left(\frac{1}{10}\right)\right], \left[\frac{1}{5}, \sin\left(\frac{1}{5}\right)\right], \left[\frac{1}{5}, 0\right]\right]$$

Define the function h to give the color of each trapezoid.

```
> h := i -> abs( cos(i/10) ):
> PLOT( seq( POLYGONS( evalf( p(i) ),
>                 COLOR(HUE, evalf( h(i) )) ),
>             i = 1..63) );
```

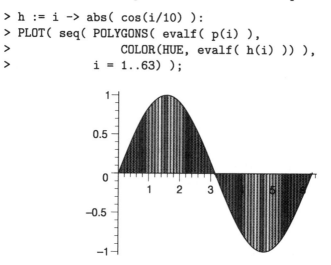

A Sum Plot

You can create procedures that directly build PLOT data structures. For example, given an unevaluated sum you can compute the partial sums, and place the values in a CURVES structure.

```
> s := Sum( 1/k^2, k=1..10 );
```

$$s := \sum_{k=1}^{10} \frac{1}{k^2}$$

You can use the typematch command to pick the unevaluated sum apart into its components.

```
> typematch( s, 'Sum'( term::algebraic,
>               n::name=a::integer..b::integer ) );
```

$$true$$

The typematch command assigns the parts of the sum to the given names.

```
> term, n, a, b;
```

$$\frac{1}{k^2}, \ k, \ 1, \ 10$$

You can now calculate the partial sums.

```
> sum( term, n=a..a+2 );
```

$$\frac{49}{36}$$

The following defines a procedure, psum, which calculates a floating-point value of the mth partial sum.

```
> psum := evalf @ unapply( Sum(term, n=a..(a+m)), m );
```

$$psum := evalf@ \left(m \to \sum_{k=1}^{1+m} \frac{1}{k^2} \right)$$

You can now create the necessary list of points.

```
> points := [ seq( [[i,psum(i)], [i+1,psum(i)]],
>     i=1..(b-a+1) ) ];
```

$$points := [[[1, 1.250000000], [2, 1.250000000]],$$
$$[[2, 1.361111111], [3, 1.361111111]],$$
$$[[3, 1.423611111], [4, 1.423611111]],$$

$$[[4, 1.463611111], [5, 1.463611111]],$$
$$[[5, 1.491388889], [6, 1.491388889]],$$
$$[[6, 1.511797052], [7, 1.511797052]],$$
$$[[7, 1.527422052], [8, 1.527422052]],$$
$$[[8, 1.539767731], [9, 1.539767731]],$$
$$[[9, 1.549767731], [10, 1.549767731]],$$
$$[[10, 1.558032194], [11, 1.558032194]]]$$

```
> points := map( op, points );
```

$points := [[1, 1.250000000], [2, 1.250000000],$
$\quad [2, 1.361111111], [3, 1.361111111], [3, 1.423611111],$
$\quad [4, 1.423611111], [4, 1.463611111], [5, 1.463611111],$
$\quad [5, 1.491388889], [6, 1.491388889], [6, 1.511797052],$
$\quad [7, 1.511797052], [7, 1.527422052], [8, 1.527422052],$
$\quad [8, 1.539767731], [9, 1.539767731], [9, 1.549767731],$
$\quad [10, 1.549767731], [10, 1.558032194],$
$\quad [11, 1.558032194]]$

This list has the right form.

```
> PLOT( CURVES( points ) );
```

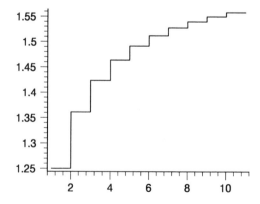

The sumplot procedure automates this technique.

```
> sumplot := proc( s )
```

```
>       local term, n, a, b, psum, m, points, i;
>       if typematch( s, 'Sum'( term::algebraic,
>            n::name=a::integer..b::integer ) ) then
>         psum := evalf @ unapply( Sum(term, n=a..(a+m)), m );
>         points := [ seq( [[i,psum(i)], [i+1,psum(i)]],
>            i=1..(b-a+1) ) ];
>         points := map(op, points);
>         PLOT( CURVES( points ) );
>       else
>         ERROR( "expecting a Sum structure as input" )
>       fi
> end:
```

Here is a sumplot of an alternating series.

```
> sumplot( Sum((-1)^k/k, k=1..25 ));
```

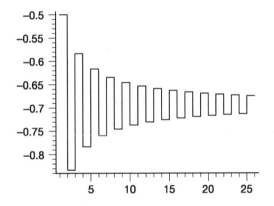

The limit of this sum is $-\ln 2$.

```
> Sum((-1)^k/k, k=1..infinity):    % = value(%);
```

$$\sum_{k=1}^{\infty} \frac{(-1)^k}{k} = -\ln(2)$$

See ?plot,structure for more details on the PLOT data structure.

The PLOT3D Data Structure

The three-dimensional plotting data structure has a form similar to the PLOT data structure. Thus, for example, the Maple expression below generates a three-dimensional plot of three lines and axes of type frame.

```
> PLOT3D( CURVES( [ [3, 3, 0], [0, 3, 1],
>                   [3, 0, 1], [3, 3, 0] ] ),
```

```
>               AXESSTYLE(FRAME) );
```

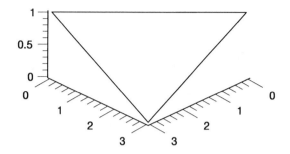

The following procedure creates the sides of a box and colors them yellow.

```
> yellowsides := proc(x, y, z, u)
>    # (x,y,0) = coordinates of a corner.
>    # z = height of box
>    # u = side length of box
>    POLYGONS(
>       [ [x,y,0], [x+u,y,0], [x+u,y,z], [x,y,z] ],
>       [ [x,y,0], [x,y+u,0], [x,y+u,z], [x,y,z] ],
>       [ [x+u, y,0], [x+u,y+u,0], [x+u,y+u,z], [x+u,y,z] ],
>       [ [x+u, y+u,0], [x,y+u,0], [x,y+u,z], [x+u,y+u,z] ],
>            COLOR(RGB,1,1,0) );
> end:
```

The redtop procedure generates a red lid for the box.

```
> redtop := proc(x, y, z, u)
>    # (x,y,z) = coordinates of a corner.
>    # u = side length of square
>    POLYGONS( [ [x,y,z], [x+u,y,z], [x+u,y+u,z], [x,y+u,z] ],
>            COLOR(RGB, 1, 0, 0) );
>   end:
```

You can now put the sides and the top inside a PLOT3D structure to display them.

```
> PLOT3D( yellowsides(1, 2, 3, 0.5),
>           redtop(1, 2, 3, 0.5),
```

```
>         STYLE(PATCH) );
```

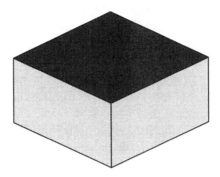

Using `yellowsides` and `redtop` you can create a three-dimensional histogram plot. Here is the histogram corresponding to $z = 1/(x + y + 4)$, for $0 \leq x \leq 4$ and $0 \leq y \leq 4$.

```
> sides := seq( seq( yellowsides(i, j, 1/(i+j+4), 0.75),
>     j=0..4), i=0..4):
> tops := seq( seq( redtop( i, j, 1/(i+j+4), 0.75),
>     j=0..4 ), i=0..4 ):
```

Histograms look nice when you enclose them in a box of axes. Axes are generated using AXESSTYLE.

```
> PLOT3D( sides, tops, STYLE(PATCH), AXESSTYLE(BOXED) );
```

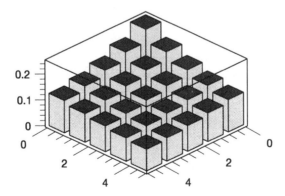

You can modify the above construction to create a `listbarchart3d` procedure which, for a given list of lists of heights, gives a three-dimensional bar chart as above for its output.

The names of the objects that can appear inside a PLOT3D data structure include all those that you can use in the PLOT data structure. Thus POINTS,

CURVES, POLYGONS, and TEXT are also available for use inside an unevaluated PLOT3D call. As in the two-dimensional case, when the object name is CURVES or POINTS, the point information consists of one or more lists of three-dimensional points, each list supplying the set of points making up a single curve in three-dimensional space. In the case of a POLYGONS structure, the point information consists of one or more lists of points. In this case, each list describes the vertices of a single polygon in three-dimensional space. There are two extra objects for PLOT3D structures. GRID is a structure that describes a functional grid. It consists of two ranges defining a grid in the x–y plane and a list of lists of z values over this grid. In the following example LL contains 4 lists each of length 3. Therefore the grid is 4×3, and x runs from 1 to 3 in increments of 2/3, whereas y runs from 1 to 2 in increments of 1/2.

```
> LL := [ [0,1,0], [1,1,1], [2,1,2], [3,0,1] ]:

> PLOT3D( GRID( 1..3, 1..2, LL ), AXESLABELS(x,y,z),
>           ORIENTATION(135, 45), AXES(BOXED) );
```

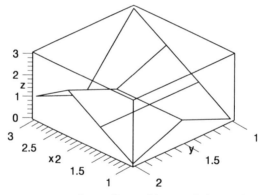

The MESH structure contains a list of lists of three-dimensional points describing a surface in three dimensions.[1]

```
> LL := [ [ [0,0,0], [1,0,0], [2,0,0], [3,0,0] ],
>         [ [0,1,0], [1,1,0], [2.1, 0.9, 0],
>                   [3.2, 0.7, 0] ],
>         [ [0,1,1], [1,1,1], [2.2, 0.6, 1],
>                   [3.5, 0.5, 1.1] ] ];
```

$$LL := [[[0, 0, 0], [1, 0, 0], [2, 0, 0], [3, 0, 0]],$$
$$[[0, 1, 0], [1, 1, 0], [2.1, .9, 0], [3.2, .7, 0]],$$
$$[[0, 1, 1], [1, 1, 1], [2.2, .6, 1], [3.5, .5, 1.1]]]$$

---

[1]An $n \times m \times 3$ hfarray is also allowed as input to MESH.

The MESH structure represents the quadrilaterals spanned by

$$LL_{i,j}, LL_{i,j+1}, LL_{i+1,j}, LL_{i+1,j+1}$$

for all meaningful values of i and j.

```
> PLOT3D( MESH( LL ), AXESLABELS(x,y,z), AXES(BOXED),
>          ORIENTATION(-140, 45) );
```

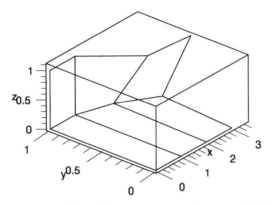

All the options available for PLOT are also available for PLOT3D. In addition, you can also use the GRIDSTYLE, LIGHTMODEL, and AMBIENTLIGHT options. See ?plot3d,structure for details on the various options to the PLOT3D structure.

8.4 Programming with Plot Data Structures

This section describes some of the tools that are available for programming at the PLOT and PLOT3D data structure level. Plotting data structures have the advantage of allowing *direct* access to all the functionality that Maple's plotting facilities provide. The examples in *Maple's Plotting Data Structures* on page 266 show the extent of the facilities' power. You could easily thicken the lines in the sum plot by adding local information to the objects in that example. This section provides a simple set of examples that describe how to program at this lower level.

Writing Graphic Primitives

You can write procedures that allow you to work with plot objects at a more conceptual level. For example, the line and disk commands in the plottools package provide a model for programming primitives such as points, lines, curves, circles, rectangles, and arbitrary polygons in both two and three dimensions. In all cases, you can specify options, such as line or patch style and color, in the same format as in other plotting procedures in Maple.

```
> line := proc(x::list, y::list)
>    # x and y represent points in either 2-D or 3-D
>    local opts;
>    opts := [ args[3..nargs] ];
>    opts := convert( opts, PLOToptions );
>    CURVES( evalf( [x, y] ), op(opts) );
> end:
```

Inside a procedure, `nargs` is the number of arguments and `args` is the actual argument sequence. Thus, in `line`, `args[3..nargs]` is the sequence of arguments that follow x and y. The `convert(..., PLOToptions)` command converts user-level options to the format that PLOT requires.

```
> convert( [axes=boxed, color=red], PLOToptions );
```

$$[AXESSTYLE(BOX),$$

$$COLOUR(RGB, 1.00000000, 0, 0)]$$

The `disk` procedure below is similar to `line` except that you can specify the number of points that `disk` should use to generate the disk. Therefore `disk` must handle that option, numpoints, separately. The `hasoption` command determines whether a certain option is present.

```
> disk := proc(x::list, r::algebraic)
>    # draw a disk of radius r centered at x in 2-D.
>    local i, n, opts, vertices;
>    opts := [ args[3..nargs] ] ;
>    if not hasoption( opts, numpoints, n, 'opts' )
>    then n := 50;
>    fi;
>    opts := convert(opts, PLOToptions);
>    vertices := seq( evalf( [ x[1] + r*cos(2*Pi*i/n),
>                              x[2] + r*sin(2*Pi*i/n) ] ),
>                     i = 0..n );
>    POLYGONS( [vertices], op(opts) );
> end:
```

You can now display two disks connected by a line as follows.

```
> with(plots):
> display( disk([-1, 0], 1/2, color=plum),
>          line([-1, 1/2], [1, 1/2]),
>          disk([1, 0], 1/2, thickness=3),
```

```
>            scaling=constrained );
```

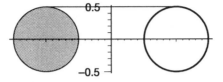

Note how the options to the individual objects apply only to those objects.

Plotting Gears

This example shows how you can manipulate plotting data structures to embed two-dimensional plots into a three-dimensional setting. The procedure below creates a little piece of the boundary of a two-dimensional graph of a gear-like structure.

```
> outside := proc(a, r, n)
>    local p1, p2;
>    p1 := evalf( [ cos(a*Pi/n), sin(a*Pi/n) ] );
>    p2 := evalf( [ cos((a+1)*Pi/n), sin((a+1)*Pi/n) ] );
>    if r = 1 then p1, p2;
>    else p1, r*p1, r*p2, p2;
>    fi
> end:
```

For example

```
> outside( Pi/4, 1.1, 16 );
```

$$[.9881327882, .1536020604],$$

$$[1.086946067, .1689622664],$$

$$[1.033097800, .3777683623],$$

$$[.9391798182, .3434257839]$$

```
> PLOT( CURVES( [%] ), SCALING(CONSTRAINED) );
```

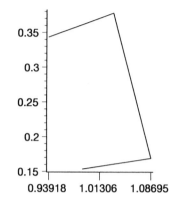

When you put the pieces together, you get a gear. SCALING(CONSTRAINED), which corresponds to the option scaling=constrained, is used to ensure that the gear appears round.

```
> points := [ seq( outside(2*a, 1.1, 16), a=0..16 ) ]:
> PLOT( CURVES(points), AXESSTYLE(NONE), SCALING(CONSTRAINED) );
```

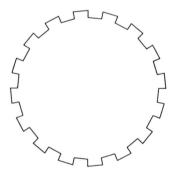

You can fill this object using the POLYGONS object. However, you must be careful, as Maple assumes that the polygons are convex. Hence, you should draw each wedge-shaped section of the gear as a triangular polygon.

```
> a := seq( [ [0, 0], outside(2*j, 1.1, 16) ], j=0..15 ):
> b := seq( [ [0, 0], outside(2*j+1, 1, 16) ], j=0..15 ):
```

```
> PLOT( POLYGONS(a,b), AXESSTYLE(NONE), SCALING(CONSTRAINED) );
```

Adding STYLE(PATCHNOGRID) to the above structure and combining it with the curve from the first picture gives you a filled gear-like structure. To embed this in three dimensions, say at a thickness of t units, you can use the utility procedures

```
> double := proc( L, t )
>    local u;
>    [ seq( [u[1], u[2], 0], u=L ) ],
>    [ seq( [u[1], u[2], t], u=L ) ];
> end:
```

which takes a list of vertices and creates two copies in three-dimensional space, one at height 0 and the second at height t, and

```
> border := proc( L1, L2 )
>    local i, n;
>    n := nops(L1);
>    seq( [ L1[i], L2[i], L2[i+1], L1[i+1] ], i = 1..n-1 ),
>       [ L1[n], L2[n], L2[1], L1[1] ];
> end:
```

which inputs two lists of vertices and joins the corresponding vertices from each list into vertices that make up quadrilaterals. You can create the top and bottom vertices of the gear embedded into three-dimensional space as follows.

```
> faces :=
> seq( double(p,1/2),
>       p=[ seq( [ outside(2*a+1, 1.1, 16), [0,0] ],
>                a=0..16 ),
>          seq( [ outside(2*a, 1,16), [0,0] ], a=0..16 )
>       ] ):
```

Now faces is a sequence of doubled outside values.

```
> PLOT3D( POLYGONS( faces ) );
```

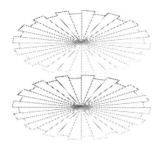

As above, the following are the points on the outline of a gear.

```
> points := [ seq( outside(2*a, 1.1, 16), a=0..16 ) ]:
> PLOT( CURVES(points), AXESSTYLE(NONE), SCALING(CONSTRAINED) );
```

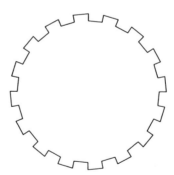

If you double these points, you get vertices of the polygons making up the border of the three-dimensional gear.

```
> bord := border( double( [ seq( outside(2*a+1, 1.1, 16),
>                                  a=0..15 ) ], 1/2) ):
```

```
> PLOT3D( seq( POLYGONS(b), b=bord ) );
```

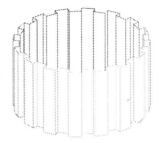

To display the gear you need to put these together in a single PLOT3D structure. Use STYLE(PATCHNOGRID) as a local option to the top and bottom of the gear so that they do not appear as several triangles.

```
> PLOT3D( POLYGONS(faces, STYLE(PATCHNOGRID) ),
>         seq( POLYGONS(b), b=bord ),
>     STYLE(PATCH), SCALING(CONSTRAINED) );
```

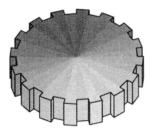

Note that the global STYLE(PATCH) and SCALING(CONSTRAINED) options apply to the whole PLOT3D structure, except where the local STYLE(PATCHNOGRID) option to the top and bottom of the gear overrides the global STYLE(PATCH) option.

Polygon Meshes

The PLOT3D *Data Structure* on page 273 describes the MESH data structure which you generate when you use plot3d to draw a parametrized surface. This simple matter involves converting a mesh of points into a set of vertices for corresponding polygons. Using polygons rather than a MESH structure allows you to modify the individual polygons. The procedure polygongrid creates the vertices of a quadrangle at the (i, j)th grid value.

```
> polygongrid := proc(gridlist, i, j)
>    gridlist[j][i], gridlist[j][i+1],
>    gridlist[j+1][i+1], gridlist[j+1][i];
> end:
```

You can then use makePolygongrid to construct the appropriate polygons.

```
> makePolygongrid := proc(gridlist)
>    local m,n,i,j;
>    n := nops(gridlist);
>    m := nops(gridlist[1]);
>    POLYGONS( seq( seq( [ polygongrid(gridlist, i, j) ],
>             i=1..m-1), j=1..n-1) );
> end:
```

The following is a mesh of points in two-dimensional space.

```
> L := [ seq( [ seq( [i-1, j-1], i=1..3 ) ], j=1..4 ) ];
```

$$L := [[[0, 0], [1, 0], [2, 0]], [[0, 1], [1, 1], [2, 1]],$$

$$[[0, 2], [1, 2], [2, 2]], [[0, 3], [1, 3], [2, 3]]]$$

The makePolygongrid procedure creates the POLYGONS structure corresponding to L.

```
> grid1 := makePolygongrid( L );
```

$$grid1 := \text{POLYGONS}([[0, 0], [1, 0], [1, 1], [0, 1]],$$

$$[[1, 0], [2, 0], [2, 1], [1, 1]], [[0, 1], [1, 1], [1, 2], [0, 2]],$$

$$[[1, 1], [2, 1], [2, 2], [1, 2]], [[0, 2], [1, 2], [1, 3], [0, 3]],$$

$$[[1, 2], [2, 2], [2, 3], [1, 3]])$$

Put the polygons inside a PLOT structure to display them.

```
> PLOT( grid1 );
```

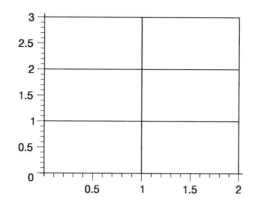

You can also use the convert(..., POLYGONS) command to convert GRID or MESH structures to polygons; see ?convert,POLYGONS. convert(..., POLYGONS) calls the procedure 'convert/POLYGONS' which, in the case of a MESH structure, works as the makePolygongrid procedure above.

8.5 Programming with the plottools Package

While the plotting data structure has the advantage of allowing direct access to all the functionality that Maple's plotting facilities provide, it does not allow you to specify colors (such as red or blue) in an intuitive way, nor does it allow you to use all the representations of numeric data, such as π or $\sqrt{2}$, that you find in Maple.

This section shows you how to work with basic graphic objects at a level higher than that of the plotting data structures. The plottools package provides commands for creating lines, disks, and other two-dimensional objects, along with commands to generate shapes such as spheres, tori, and polyhedra. For example, one can draw a sphere of unit radius and a torus at specified center using a patch style of rendering and a frame style of axis.

```
> with(plots): with(plottools):
> display( sphere( [0, 0, 2] ), torus( [0, 0, 0] ),
```

```
>                    style=patch, axes=frame, scaling=constrained );
```

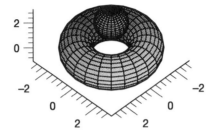

and rotate it at various angles via the functions in the plottools package.

```
> rotate( %, Pi/4, -Pi/4, Pi/4 );
```

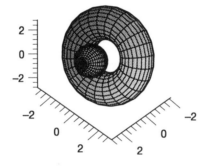

A Pie Chart

You can write a plotting procedure to build a pie chart of a list of integer data. The piechart procedure below uses the following partialsum procedure which calculates the partial sums of a list of numbers up to a given term.

```
> partialsum := proc(d, i)
>    local j;
>    evalf( Sum( d[j], j=1..i ) )
> end:
```

For example

```
> partialsum( [1, 2, 3, -6], 3 );
```

$$6.$$

The piechart procedure first computes the relative weights of the data along with the centers of each pie slice. piechart uses a TEXT structure to place the data information at the center of each pie slice and the pieslice command from the plottools package to generate the pie slices. Finally, piechart also varies the colors of each slice by defining a color function based on hue coloring.

```
> piechart := proc( data::list(integer) )
>    local b, c, i, n, x, y, total;
>
>    n := nops(data);
>    total := partialsum(data, n);
>    b := 0, seq( evalf( 2*Pi*partialsum(data, i)/total ),
>                 i =1..n );
>    x := seq( ( cos(b[i])+cos(b[i+1]) ) / 3, i=1..n ):
>    y := seq( ( sin(b[i])+sin(b[i+1]) ) / 3, i=1..n ):
>    c := (i, n) -> COLOR(HUE, i/(n + 1)):
>    PLOT( seq( plottools[pieslice]( [0, 0], 1,
>                     b[i]..b[i+1], color=c(i, n) ),
>             i=1..n),
>          seq( TEXT( [x[i], y[i]],
>                     convert(data[i], name) ),
>               i = 1..n ),
>          AXESSTYLE(NONE), SCALING(CONSTRAINED) );
> end:
```

Here is a piechart with six slices.

```
> piechart( [ 8, 10, 15, 10, 12, 16 ] );
```

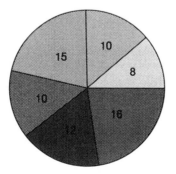

The AXESSTYLE(NONE) option ensures that Maple does not draw any axes with the pie chart.

A Dropshadow Procedure

You can use the existing procedures to create other types of plots that are not part of the available Maple graphics library. For example, the following procedure computes the three-dimensional plot of a surface, $z = f(x, y)$, that has a dropshadow projection onto a plane located below the surface. The procedure makes use of the commands `contourplot`, `contourplot3d`, `display` from the `plots` package, and `transform` from the `plottools` package.

```
> dropshadowplot := proc(F::algebraic, r1::name=range,
>       r2::name=range, r3::name=range)
>    local minz, p2, p3, coption, opts, f, g, x, y;
>
>    # set the number of contours (default 8)
>    opts := [args[5..nargs]];
>    if not hasoption( opts, 'contours', coption, 'opts' )
>    then coption := 8;
>    fi;
>
>    # determine the base of the plot axes
>    # from the third argument
>    minz := lhs('if'(r3::range, r3, rhs(r3)));
>    minz := evalf(minz);
>
>
>    # create 2d and 3d contour plots for F.
>    p3 := plots[contourplot3d]( F, r1, r2,
>             'contours'=coption, op(opts) );
>    p2 := plots[contourplot]( F, r1, r2,
>             'contours'=coption, op(opts) );
>
>    # embed contour plot into R^3 via plottools[transform]
>    g := unapply( [x,y,minz], x, y );
>    f := plottools[transform]( g );
>    plots[display]([ f(p2), p3 ]);
> end:
```

The `filled=true` option to `contourplot` and `contourplot3d` causes these two commands to fill the regions between the level curves with a color that indicates the level.

```
> expr := -5 * x / (x^2+y^2+1);
```

$$expr := -5\,\frac{x}{x^2 + y^2 + 1}$$

```
> dropshadowplot( expr, x=-3..3, y=-3..3, z=-4..3,
>     filled=true, contours=3, axes=frame );
```

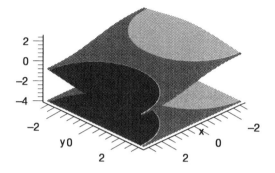

The first section of the dropshadow procedure determines if you have specified a contours option in the optional arguments (those after the fourth argument), making use of the hasoption procedure. The next section of dropshadowplot determines the z value of the base. Note that you must take care since you specify ranges differently for formula or function input. The remaining sections create the correct plotting objects which represent the two types of contour plots. dropshadowplot embeds the two-dimensional contour plot into three-dimensional space using the transformation

$$(x, y) \mapsto [x, y, minz]$$

going from $R^2 \rightarrow R^3$. Finally, it displays the two plots together in one three-dimensional plotting object.

Note that you can either provide an alternate number of levels or even specify the precise contour locations via the contours option. Thus,

```
> dropshadowplot( expr, x=-3..3, y=-3..3, z=-4..3,
>                 filled=true, contours=[-2,-1,0,1,2] );
```

produces a plot similar to that shown above, except now it produces 5 contours at levels $-2, -1, 0, 1$, and 2.

Creating a Tiling

The plottools package provides a convenient environment for programming graphical procedures. For example, you can draw circular arcs in a unit square.

```
> with(plots): with(plottools):
```

```
> a := rectangle( [0,0], [1,1] ),
>       arc( [0,0], 0.5, 0..Pi/2 ),
>       arc( [1,1], 0.5, Pi..3*Pi/2 ):
> b := rectangle( [1.5,0], [2.5,1] ),
>       arc( [1.5,1], 0.5, -Pi/2..0 ),
>       arc( [2.5,0], 0.5, Pi/2..Pi ):
```

You must use display from plots to show the objects that rectangle and arc create.

```
> display( a, b, axes=none, scaling=constrained );
```

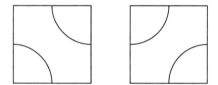

You can tile the plane with a and b type rectangles. The following procedure creates such a $m \times n$ tiling using a function, g, to determine when to use an a-tile and when to use a b-tile. The function g should return either 0, to use an a-tile, or 1, to use a b-tile.

```
> tiling := proc(g, m, n)
>    local i, j, r, h, boundary, tiles;
>
>    # define an a-tile
>    r[0] := plottools[arc]( [0,0], 0.5, 0..Pi/2 ),
>            plottools[arc]( [1,1], 0.5, Pi..3*Pi/2 );
>    # define a b-tile
>    r[1] := plottools[arc]( [0,1], 0.5, -Pi/2..0 ),
>            plottools[arc]( [1,0], 0.5, Pi/2..Pi );
>    boundary := plottools[curve]( [ [0,0], [0,n],
>                [m,n], [m,0], [0,0]] );
>    tiles := seq( seq( seq( plottools[translate](h, i, j),
>            h=r[g(i, j)] ), i=0..m-1 ), j=0..n-1 );
>    plots[display]( tiles, boundary, args[4..nargs] );
> end:
```

As an example, define the following procedure which randomly returns either 0 or 1.

```
> oddeven := proc() rand() mod 2 end:
```

Create a 20×10 tiling (called a Truchet tiling) with no axes and constrained scaling.

```
> tiling( oddeven, 20, 10, scaling=constrained, axes=none);
```

When you use the same procedure again, the random tiling is different.

```
> tiling( oddeven, 20, 10, scaling=constrained, axes=none);
```

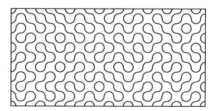

A Smith Chart

The commands in the `plottools` package allow for easy creation of such useful graphs as a Smith Chart, used in microwave circuit analysis.

```
> smithChart := proc(r)
>     local i, a, b, c ;
>     a := PLOT( seq( plottools[arc]( [-i*r/4,0],
>                                      i*r/4, 0..Pi ),
>                     i = 1..4 ),
```

```
>        plottools[arc]( [0,r/2], r/2,
>                             Pi-arcsin(3/5)..3*Pi/2 ),
>        plottools[arc]( [0,r], r, Pi..Pi+arcsin(15/17) ),
>        plottools[arc]( [0,2*r], 2*r,
>                             Pi+arcsin(3/5)..Pi+arcsin(63/65) ),
>        plottools[arc]( [0,4*r], 4*r,
>                             Pi+arcsin(15/17)..Pi+arcsin(63/65) )
>              );
>    b := plottools[transform]( (x, y) -> [x,-y] )(a);
>    c := plottools[line]( [ 0, 0], [ -2*r, 0] ):
>    plots[display]( a, b, c, axes = none,
>                    scaling = constrained,
>                    args[2..nargs] );
> end:
```

Here is a Smith Chart of radius 1.

```
> smithChart( 1 );
```

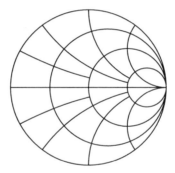

Make a Smith Chart by building appropriate circular arcs above the axes, creating a copy reflected on the axis (using the transform procedure), and then adding a final horizontal line. The parameter r denotes the radius of the largest circle. Modifying the smithChart procedure to add text to mark appropriate grid markers is a simple operation.

Modifying Polygon Meshes

You can easily construct a new plot tool that works like those in the plottools package. For example, you can cut out or modify polygon structures by first working with individual faces and then mapping the results onto entire polygons. Thus, you can have a procedure that cuts out the inside of a single face of a polygon.

```
> cutoutPolygon := proc( vlist_in::{list, hfarray},
```

```
>                          scale::numeric )
>    local vlist, i, center, outside, inside, n, edges, polys;
>
>    vlist := 'if'(vlist_in::hfarray, op(3, eval(vlist_in)),
>                  vlist_in);
>    n := nops(vlist);
>    center := add( i, i=vlist ) / n;
>    inside := seq( scale*(vlist[i]-center) + center,
>                   i=1..n);
>    outside := seq( [ inside[i],  vlist[i],
>                      vlist[i+1], inside[i+1] ],
>                   i=1..n-1 ):
>    polys := POLYGONS( outside,
>                       [ inside[n], vlist[n],
>                         vlist[1], inside[1] ],
>                       STYLE(PATCHNOGRID) );
>    edges := CURVES( [ op(vlist), vlist[1] ],
>                     [ inside, inside[1] ] );
>    polys, edges;
> end:
```

Note that cutoutPolygon was written to handle input in either standard matrix form or hfarray form.

The following are the corners of a triangle.

```
> triangle := [ [0,2], [2,2], [1,0] ];
```

$$triangle := [[0, 2], [2, 2], [1, 0]]$$

The cutoutPolygon procedure converts triangle to three polygons (one for each side) and two curves.

```
> cutoutPolygon( triangle, 1/2 );
```

$$\text{POLYGONS}\left(\left[\left[\frac{1}{2}, \frac{5}{3}\right], [0, 2], [2, 2], \left[\frac{3}{2}, \frac{5}{3}\right]\right],\right.$$

$$\left[\left[\frac{3}{2}, \frac{5}{3}\right], [2, 2], [1, 0], \left[1, \frac{2}{3}\right]\right], \left[\left[1, \frac{2}{3}\right], [1, 0], [0, 2], \left[\frac{1}{2}, \frac{5}{3}\right]\right],$$

$$\text{STYLE}(PATCHNOGRID)), \text{CURVES}($$

$$[[0, 2], [2, 2], [1, 0], [0, 2]], \left[\left[\frac{1}{2}, \frac{5}{3}\right], \left[\frac{3}{2}, \frac{5}{3}\right], \left[1, \frac{2}{3}\right], \left[\frac{1}{2}, \frac{5}{3}\right]\right]\right)$$

Use the display command from the plots package to show the triangle.

```
> plots[display]( %, color=red );
```

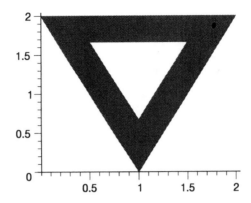

The cutout procedure below applies cutoutPolygon to every face of a polyhedron.

```
> cutout := proc(polyhedron, scale)
>    local v;
>    seq( cutoutPolygon( v, evalf(scale) ), v=polyhedron);
> end:
```

You can now cut out 3/4 of each face of a dodecahedron.

```
> display( cutout( dodecahedron([1, 2, 3]), 3/4 ),
>          scaling=constrained);
```

As a second example, you can take a polygon and raise or lower its barycenter.

```
> stellateFace := proc( vlist::list, aspectRatio::numeric )
>    local apex, i, n;
>
>    n := nops(vlist);
```

```
>     apex :=  add( i, i = vlist ) * aspectRatio / n;
>     POLYGONS( seq( [ apex, vlist[i],
>                        vlist[modp(i, n) + 1] ],
>                     i=1..n) );
> end:
```

The following are the corners of a triangle in three-dimensional space.

```
> triangle := [ [1,0,0], [0,1,0], [0,0,1] ];
```

$$triangle := [[1, 0, 0], [0, 1, 0], [0, 0, 1]]$$

The stellateFace procedure creates three polygons, one for each side of the triangle.

```
> stellateFace( triangle, 1 );
```

$$\text{POLYGONS}\left(\left[\left[\frac{1}{3}, \frac{1}{3}, \frac{1}{3}\right], [1, 0, 0], [0, 1, 0]\right],\right.$$

$$\left[\left[\frac{1}{3}, \frac{1}{3}, \frac{1}{3}\right], [0, 1, 0], [0, 0, 1]\right],$$

$$\left.\left[\left[\frac{1}{3}, \frac{1}{3}, \frac{1}{3}\right], [0, 0, 1], [1, 0, 0]\right]\right)$$

Since these POLYGONS belong in three-dimensional space, you must put them inside a PLOT3D structure to display them.

```
> PLOT3D( % );
```

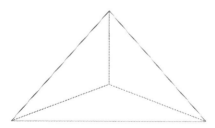

Again, you can extend stellateFace to work for arbitrary polyhedra having more than one face.

```
> stellate := proc( polyhedron, aspectRatio)
>     local v;
```

```
>    seq( stellateFace( v, evalf(aspectRatio) ),
>        v=polyhedron );
> end:
```

This allows for the construction of stellated polyhedra.

```
> stellated := display( stellate( dodecahedron(), 3),
>        scaling= constrained ):
> display( array( [dodecahedron(), stellated] ) );
```

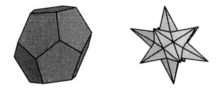

You can use convert(..., POLYGONS) to convert a GRID or MESH structure to the equivalent set of POLYGONS. Here is a POLYGONS version of the Klein bottle.

```
> kleinpoints := proc()
>    local bottom, middle, handle, top, p, q;
>
>    top := [ (2.5 + 1.5*cos(v)) * cos(u),
>            (2.5 + 1.5*cos(v)) * sin(u), -2.5 * sin(v) ]:
>    middle := [ (2.5 + 1.5*cos(v)) * cos(u),
>            (2.5 + 1.5*cos(v)) * sin(u), 3*v - 6*Pi ]:
>    handle := [ 2 - 2*cos(v) + sin(u), cos(u),
>            3*v - 6*Pi ]:
>    bottom := [ 2 + (2+cos(u))*cos(v), sin(u),
>            -3*Pi + (2+cos(u)) * sin(v) ]:
>    p := plot3d( {bottom, middle, handle, top},
>            u=0..2*Pi, v=Pi..2*Pi, grid=[9,9] ):
>    p := select( x -> op(0,x)=MESH, [op(p)] );
>    seq( convert(q , POLYGONS), q=p );
> end:
> display( kleinpoints(), style=patch,
```

```
>              scaling=constrained, orientation=[-110,71] );
```

You can then use the commands for manipulation of polygons to view the inside of the Klein bottle above.

```
> display( seq( cutout(k, 3/4), k=kleinpoints() ),
>              scaling=constrained );
```

8.6 Example: Vector Field Plots

This section describes the problem of plotting a vector field of two dimensional vectors in the plane. The example herein serves to pinpoint some of tools available for plot objects on grids in two- and three-dimensional space.

The command to plot a vector field should have the following syntax.

```
vectorfieldplot( F, r1, r2 , options )
```

The input, *F*, is a list of size two, giving the functions that make up the horizontal and vertical components of the vector field. The arguments *r1* and *r2* describe the domain grid of the vectors. The three arguments *F*, *r1*, and *r2* are similar in form to the input you need to use for plot3d. Similarly, the optional information includes any sensible specification that plot or plot3d allows. Thus, options of the form grid = [*m*,*n*], style = patch, and color = *colorfunction* are valid options.

The first problem is to draw a vector. Let [*x*, *y*] represent a point, the starting point of the arrow, and [*a*, *b*], the components of the vector. You can determine the shape of an arrow by three independent parameters, *t*1, *t*2, and *t*3. Here *t*1 denotes the thickness of the arrow, *t*2 the thickness of the arrow head, and *t*3 the ratio of the length of the arrow head in comparison to the length of the arrow itself.

The procedure arrow below from the plottools package constructs seven vertices of an arrow. It then builds the arrow by constructing two polygons: a triangle (spanned by v_5, v_6, and v_7) for the head of the arrow and a rectangle (spanned by v_1, v_2, v_3, and v_4) for the tail; it then removes boundary lines by setting the style option inside the polygon structure. It also constructs the boundary of the entire arrow via a closed curve through the vertices.

```
> arrow := proc( point::list, vect::list, t1, t2, t3)
>    local a, b, i, x, y, L, Cos, Sin, v, locopts;
>
>    a := vect[1]; b := vect[2];
>    if has( vect, 'undefined') or (a=0 and b=0) then
>       RETURN( POLYGONS( [ ] ) );
>    fi;
>    x := point[1]; y := point[2];
>    # L = length of arrow
>    L := evalf( sqrt(a^2 + b^2) );
>    Cos := evalf( a / L );
>    Sin := evalf( b / L);
>    v[1] := [x + t1*Sin/2, y - t1*Cos/2];
>    v[2] := [x - t1*Sin/2, y + t1*Cos/2];
>    v[3] := [x - t1*Sin/2 - t3*Cos*L + a,
>             y + t1*Cos/2 - t3*Sin*L + b];
>    v[4] := [x + t1*Sin/2 - t3*Cos*L + a,
>             y - t1*Cos/2 - t3*Sin*L + b];
>    v[5] := [x - t2*Sin/2 - t3*Cos*L + a,
>             y + t2*Cos/2 - t3*Sin*L + b];
>    v[6] := [x + a, y + b];
>    v[7] := [x + t2*Sin/2 - t3*Cos*L + a,
>             y - t2*Cos/2 - t3*Sin*L + b];
>    v := seq( evalf(v[i]), i= 1..7  );
```

```
>
>      # convert optional arguments to PLOT data structure form
>      locopts := convert( [style=patchnogrid,
>                           args[ 6..nargs ] ],
>                         PLOToptions );
>      POLYGONS( [v[1], v[2], v[3], v[4]],
>                [v[5], v[6], v[7]], op(locopts) ),
>      CURVES( [v[1], v[2], v[3], v[5], v[6],
>              v[7], v[4], v[1]] );
> end:
```

Note that you must build the polygon structure for the arrow in two parts, because each polygon must be convex. In the special case where the vector has both components equal to zero or an undefined component, such as a value resulting from a non-numeric value (for example, a complex value or a singularity point), the arrow procedure returns a trivial polygon. Here are four arrows.

```
> arrow1 := PLOT(arrow( [0,0], [1,1], 0.2, 0.4, 1/3,
>                 color=red) ):
> arrow2 := PLOT(arrow( [0,0], [1,1], 0.1, 0.2, 1/3,
>                 color=yellow) ):
> arrow3 := PLOT(arrow( [0,0], [1,1], 0.2, 0.3, 1/2,
>                 color=blue) ):
> arrow4 := PLOT(arrow( [0,0], [1,1], 0.1, 0.5, 1/4,
>                 color=green) ):
```

The display command from the plots package can show an array of plots.

```
> with(plots):

> display( array( [[arrow1, arrow2], [arrow3, arrow4 ]] ),
>     scaling=constrained );
```

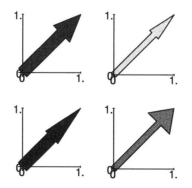

The remainder of this section presents a number of solutions to the programming problem of generating a vector field plot, each a bit more powerful than its predecessors. The first and simplest solution requires the input to be in functional (rather than expression) form. You first need three utility procedures that process the domain information, generate a grid of function values, and place the information in a PLOT3D structure.

The procedure `domaininfo` determines the endpoints and increments for the grid. `domaininfo` takes as input the two ranges `r1` and `r2` and the two grid sizes m and n, and returns the grid information as an expression sequence of four elements.

```
> domaininfo := proc(r1, r2, m, n)
>     lhs(r1), lhs(r2),
>     evalf( (rhs(r1) - lhs(r1))/(m-1) ),
>     evalf( (rhs(r2) - lhs(r2))/(n-1) );
> end:
```

Here is an example using multiple assignments to assign the four values returned to separate variables.

```
> a, b, dx, dy := domaininfo( 0..12, 20..100, 7, 9);
```

$$a, b, dx, dy := 0, 20, 2., 10.$$

Now a, b, dx, and dy have the following values.

```
> a, b, dx, dy;
```

$$0, 20, 2., 10.$$

For the conversion to a grid of numerical points, you can take advantage of the extendibility of Maple's convert command. The procedure `convert/grid` below takes a function f as input and evaluates it over the grid which `r1`, `r2`, `m`, and `n` specify.

```
> 'convert/grid' := proc(f, r1, r2, m, n)
>     local a, b, i, j, dx, dy;
>     # obtain  information about domain
>     a,b,dx,dy := domaininfo( r1, r2, m, n );
>     # output grid of function values
>     [ seq( [ seq( evalf( f( a + i*dx, b + j*dy ) ),
>         i=0..(m-1) ) ], j=0..(n-1) ) ];
> end:
```

Now you can evaluate the undefined function, f, on a grid as follows.

```
> convert( f, grid, 1..2, 4..6, 3, 2 );
```

$$[[f(1, 4), f(1.500000000, 4), f(2.000000000, 4)],$$

$$[f(1, 6.), f(1.500000000, 6.), f(2.000000000, 6.)]]$$

The final utility procedure determines the scalings which ensure that the arrows do not overlap. Then generateplot calls upon the arrow procedure to draw the vectors. Note that generateplot moves the origin of each arrow to center it over its grid-point.

```
> generateplot := proc(vect1, vect2, m, n, a, b, dx, dy)
>     local i, j, L, xscale, yscale, mscale;
>
>     # Determine scaling factors.
>     L := max( seq( seq( vect1[j][i]^2 + vect2[j][i]^2,
>               i=1..m ), j=1..n ) );
>     xscale := evalf( dx/2/L^(1/2) );
>     yscale := evalf( dy/2/L^(1/2) );
>     mscale := max(xscale, yscale);
>
>     # Generate plot data structure.
>     # Each arrow is centered over its point.
>     PLOT( seq( seq( arrow(
>        [ a + (i-1)*dx - vect1[j][i]*xscale/2,
>          b + (j-1)*dy - vect2[j][i]*yscale/2 ],
>        [ vect1[j][i]*xscale, vect2[j][i]*yscale ],
>          mscale/4, mscale/2, 1/3 ), i=1..m), j=1..n) );
>     # Thickness of tail = mscale/4
>     # Thickness of head = mscale/2
> end:
```

With these utility functions in place, you are ready to make the first vectorfieldplot command by putting them all together.

```
> vectorfieldplot := proc(F, r1, r2, m, n)
>     local vect1, vect2, a, b, dx, dy;
>
>     # Generate each component over the grid of points.
>     vect1 := convert( F[1], grid, r1, r2 ,m, n );
>     vect2 := convert( F[2], grid, r1, r2 ,m, n );
>
>     # Obtain the domain grid information from r1 and r2.
>     a,b,dx,dy := domaininfo(r1, r2, m, n);
>
>     # Generate the final plot structure.
>     generateplot(vect1, vect2, m, n, a, b, dx, dy)
> end:
```

Try the procedure out on the vector field $(\cos(xy), \sin(xy))$.

```
> p := (x,y) -> cos(x*y): q := (x,y) -> sin(x*y):
```

```
> vectorfieldplot( [p, q], 0..Pi, 0..Pi, 15, 20 );
```

The vectorfieldplot code shows how to write a procedure that generates vector field plots based on alternative descriptions of the input. For example, you could create a procedure listvectorfieldplot, with the input consisting of a list of *m* lists, each of which consists of *n* pairs of points. Each pair of points represents the components of a vector. The domain grid would be $1, \ldots, m$ in the horizontal direction and $1, \ldots, n$ in the vertical direction (as for listplot3d from the plots package).

```
> listvectorfieldplot := proc(F)
>     local m, n, vect1, vect2;
>
>     n := nops( F );  m := nops( F[1] );
>     # Generate the 1st and 2nd components  of F.
>     vect1 := map( u -> map( v -> evalf(v[1]) , u) , F);
>     vect2 := map( u -> map( v -> evalf(v[2]) , u) , F);
>
>     # Generate the final plot structure.
>     generateplot(vect1, vect2, m, n, 1, 1, m-1, n-1)
> end:
```

For example, the list

```
> l := [ [ [1,1], [2,2], [3,3] ],
>         [ [1,6], [2,0], [5,1] ] ]:
```

plots as

```
> listvectorfieldplot( l );
```

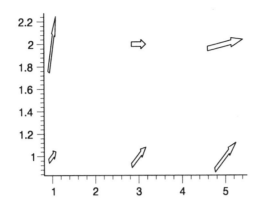

At this stage, the `vectorfieldplot` procedure still has problems. The first is that the procedure only works with function input, rather than with both function and formula input. You can solve this by converting formula expressions to procedures, and then have `vectorfieldplot` call itself recursively with the new output as in the `ribbonplot` procedure in *A Ribbon Plot Procedure* on page 263.

A second problem occurs with input like the following.

```
> p  := (x,y) -> x*y:
> q1 := (x,y) -> 1/sin(x*y):
> q2 := (x,y) -> log(x*y):
> q3 := (x,y) -> log(-x*y):
```

Maple generates an error message, since it divides by zero at points like $(0, 0)$.

```
> vectorfieldplot( [p, q1], 0..1, 0..Pi, 15, 20);
```

```
Error, (in q1) division by zero
```

Here Maple encounters the singularity $\ln(0)$.

```
> vectorfieldplot( [p, q2], 0..1, 0..Pi, 15, 20);
```

```
Error, (in ln) singularity encountered
```

In this last example, $\ln(-xy)$ is a complex number, and so finding the maximum of the function's values is meaningless.

```
> vectorfieldplot( [p, q3], 1..2, -2..1, 15, 20);
```

```
Error, (in simpl/max) constants must be real
```

A third problem is that `vectorfieldplot` only works with lists as input, not hfarrays.

To overcome such problems, ensure that you first convert all input functions to functions that only output a numeric real value or the value undefined, the only type of data the Maple plotting data structure accepts. You may also want to use the more efficient hardware floating-point calculations rather than software floating-point operations, whenever possible. *Generating Grids of Points* on page 309 describes how to do this. Instead of writing your own procedure for computing the grid, you can use the library function convert(..., gridpoints) which, in the case of a single input, generates a structure of the following form.

```
[ a..b, c..d, [ [z11, ..., z1n], ...,
[ zm1 , ..., zmn ] ] ]
```

The third argument may also be an hfarray.

It uses either expressions or procedures as input. The output gives the domain information $a..b$ and $c..d$ along with the z values of the input that it evaluates over the grid.

```
> convert( sin(x*y), 'gridpoints',
>     x=0..Pi, y=0..Pi, grid=[2, 3] );
```

$$[0..3.14159265358979, 0..3.14159265358979, [$$

$$[0, 0, 0],$$

$$[0, -.975367972083633572, -.430301217000074065]]]$$

When $xy > 0$ and $\ln(-xy)$ is complex, the grid contains the value undefined.

```
> convert( (x,y) -> log(-x*y), 'gridpoints',
>     1..2, -2..1,  grid=[2,3] );
```

$$[1...2., -2...1., [[.693147180559945286,$$

$$-.693147180559945286, \textit{undefined}],$$

$$[1.386294361, 0, \textit{undefined}]]]$$

The version of vectorfieldplot below makes use of the convert(..., gridpoints) procedure. The vectorfieldplot command should allow a number of options. In particular, it should allow a grid = $[m,n]$ option. You can accomplish this by passing the options to convert(..., gridpoints). The utility procedure makevectors handles the interface to convert(..., gridpoints).

```
> makevectors := proc( F, r1, r2  )
>     local v1, v2;
```

```
>
>      # Generate the numerical grid
>      # of components of the vectors.
>      v1 := convert( F[1], 'gridpoints', r1, r2,
>                     args[4 .. nargs] );
>      v2 := convert( F[2], 'gridpoints', r1, r2,
>                     args[4 .. nargs] );
>
>      # The domain information is contained in first
>      # two operands of v1. The function values in
>      # the 3rd components of v1 and v2.
>      [ v1[1], v1[2], v1[3], v2[3] ]
> end:
```

Here is the new version of vectorfieldplot.

```
> vectorfieldplot := proc(F, r1, r2)
>    local R1, R2, m, n, a, b, v1, v2, dx, dy, v;
>
>    v := makevectors( F, r1, r2, args[4..nargs] );
>    R1 := v[1];  R2 := v[2];  v1 := v[3];  v2 := v[4];
>
>    n := nops(v1); m := nops(v1[1]);
>    a,b,dx,dy := domaininfo(R1, R2, m, n);
>
>    generateplot(v1, v2, m, n, a, b, dx, dy);
> end:
```

Test this procedure.

```
> p := (x,y) -> cos(x*y):
> q := (x,y) -> sin(x*y):
> vectorfieldplot( [p, q], 0..Pi, 0..Pi,
>    grid=[3, 4] );
```

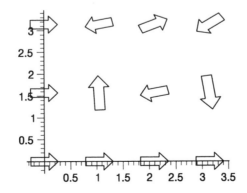

All the versions of `vectorfieldplot` so far have scaled each arrow so that each vector fits into a single grid box. No overlapping of arrows occurs. However, the arrows still vary in length. Often this results in graphs that have a large number of very small, almost invisible vectors. For example, a plot of the gradient field of $F = \cos(xy)$ exhibits this behavior.

```
> vectorfieldplot( [y*cos(x*y), x*sin(x*y)],
>     x=0..Pi, y=0..Pi, grid=[15,20]);
```

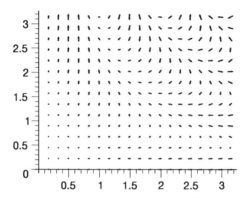

The final version of `vectorfieldplot` differs in that all the arrows have the same length—the color of each vector provides the information about the magnitudes of the arrows. You must add a utility procedure that generates a grid of colors from the function values.

```
> 'convert/colorgrid' := proc( colorFunction )
>     local colorinfo, i, j, m, n;
>
>     colorinfo := op( 3, convert(colorFunction,
>         'gridpoints', args[2..nargs] ) );
>     map( x -> map( y -> COLOR(HUE, y), x) , colorinfo );
> end:
```

The above procedure uses the `convert( ... , gridpoints)` to generate a list of lists of function values that specify the colors (using hue coloring).

```
> convert( sin(x*y), 'colorgrid',
>             x=0..1, y=0..1, grid=[2,3] );
```

$$[[$$
$$COLOR(\textit{HUE}, 0), \; COLOR(\textit{HUE}, 0), \; COLOR(\textit{HUE}, 0)$$
$$], [COLOR(\textit{HUE}, 0),$$
$$COLOR(\textit{HUE}, .479425538604203005),$$
$$COLOR(\textit{HUE}, .841470984807896505)]]$$

Here is the final version of vectorfieldplot.

```
> vectorfieldplot := proc( F, r1, r2 )
>    local v, m, n, a, b, dx, dy, opts, p, v1, v2,
>          L, i, j, norms, colorinfo,
>          xscale, yscale, mscale;
>
>    v := makevectors( F, r1, r2, args[4..nargs] );
>    v1 := v[3];   v2 := v[4];
>    n := nops(v1); m := nops( v1[1] );
>
>    a,b,dx,dy := domaininfo(v[1], v[2], m, n);
>
>    # Determine the function used for coloring the arrows.
>    opts := [ args[ 4..nargs] ];
>    if not hasoption( opts, color, colorinfo, 'opts' ) then
>        # Default coloring will be via
>        # the scaled magnitude of the vectors.
>        L := max( seq( seq( v1[j][i]^2 + v2[j][i]^2,
>                  i=1..m ), j=1..n ) );
>        colorinfo := ( F[1]^2 + F[2]^2 )/L;
>    fi;
>
>    # Generate the information needed to color the arrows.
>    colorinfo := convert( colorinfo, 'colorgrid',
>          r1, r2, op(opts) );
>
>    # Get all the norms of the vectors using zip.
>    norms := zip( (x,y) -> zip( (u,v)->
>       if u=0 and v=0 then 1 else sqrt(u^2 + v^2) fi,
>          x, y), v1, v2);
>    #  Normalize v1 and v2 (again using zip ).
>    v1 := zip( (x,y) -> zip( (u,v)-> u/v, x, y),
>          v1, norms );
>
>    v2 := zip( (x,y) -> zip( (u,v)-> u/v, x, y),
>          v2, norms );
>
>    # Generate scaling information and plot data structure.
>    xscale := dx/2.0;   yscale := dy/2.0;
>    mscale := max(xscale, yscale);
>
>    PLOT( seq( seq( arrow(
>          [ a + (i-1)*dx - v1[j][i]*xscale/2,
>            b + (j-1)*dy - v2[j][i]*yscale/2 ],
>          [ v1[j][i]*xscale, v2[j][i]*yscale ],
```

```
>            mscale/4, mscale/2, 1/3,
>            'color'=colorinfo[j][i]
>                      ), i=1..m ), j=1..n ) );
> end:
```

With this new version you can obtain the following plots.

```
> vectorfieldplot( [y*cos(x*y), x*sin(x*y)],
>    x=0..Pi, y=0..Pi,grid=[15,20] );
```

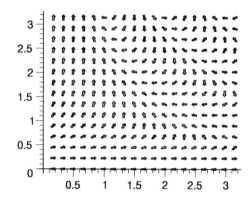

You can color the vectors via a function, such as $\sin(xy)$.

```
> vectorfieldplot( [y*cos(x*y), x*sin(x*y)],
>    x=0..Pi, y=0..Pi, grid=[15,20], color=sin(x*y) );
```

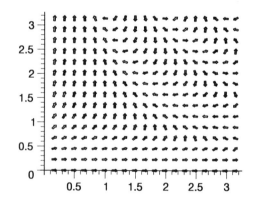

Other vector field routines can be derived from the routines above. For example, you can also write a complex vector field plot that takes complex number locations and complex number descriptions of vectors as input. You simply need to generate the grid of points in an alternate manner.

8.7 Generating Grids of Points

Example: Vector Field Plots on page 297 points out that the simple operation of obtaining an array of grid values for a given procedure, that is, the problem of computing the values of a function you wish to plot over a grid of points, is not an obvious task. You must deal with efficiency, error conditions, and non-numeric output (such as complex numbers). You can handle the case where the input is a formula in two variables in the same way as in the `ribbonplot` procedure from *A Ribbon Plot Procedure* on page 263. Thus, for simplicity of presentation, this section avoids this particular case.

The goal is to compute an array of values for f at each point on a $m \times n$ rectangular grid. That is, at the locations

$$x_i = a + (i-1)\delta_x \quad \text{and} \quad y_j = c + (j-1)\delta_y$$

where $\delta_x = (b-a)/(m-1)$ and $\delta_y = (d-c)/(n-1)$. Here i and j vary from 1 to m and 1 to n, respectively.

Consider the function $f: (x, y) \mapsto 1/\sin(xy)$. You need to evaluate f over the $m \times n$ grid with the ranges $a, \ldots, b$ and $c, \ldots, d$.

```
> f := (x,y) -> 1 / sin(x*y);
```

$$f := (x, y) \to \frac{1}{\sin(x\,y)}$$

The first step is to convert the function f to a numeric procedure. Since Maple requires numeric values (rather than symbolic) for plots, ask Maple to convert f to a form which returns numerical answers or the special value `undefined`.

```
> fnum := convert( f , numericproc );
```

$fnum :=$ **proc**($\_X$, $\_Y$)

   local *err*;

      err := traperror(evalhf(f($\_X$, $\_Y$)));

      if type([*err*], [*numeric*]) **then** *err*

      else

         err := traperror(evalf(f($\_X$, $\_Y$)));

         if type([*err*], [*numeric*]) **then** *err* **else** *undefined* **fi**

   fi

     end

The above procedure, which is the result of this conversion, attempts to calculate the numerical values as efficiently as possible. Hardware floating-point arithmetic, although of limited precision, is more efficient than software floating-point and is frequently sufficient for plotting. Thus, fnum tries evalhf first. If evalhf is successful, it returns a numeric result; otherwise, it generates an error message. If this happens, fnum attempts the calculation again using software floating-point arithmetic by calling evalf. Even this calculation is not always possible. In the case of f, the function is undefined whenever $x = 0$ or $y = 0$. In such cases, the procedure fnum returns the name undefined. Maple's plot display routines recognize this special name.

    At the point $(1, 1)$, the function f has the value $1/\sin(1)$ and so fnum returns a numerical estimate.

```
> fnum(1,1);
```

$$1.18839510577812124$$

However, if you instead try to evaluate this same function at $(0, 0)$, Maple informs you that the function is undefined at these coordinates.

```
> fnum(0,0);
```

undefined

Creating such a procedure is the first step in creating the grid of values.

    For reasons of efficiency, you should, whenever you can, compute not only the function values but also the grid points using hardware floating-point arithmetic. In addition, you should do as much computation as possible in a single call to evalhf. Whenever you use hardware floating-point arithmetic, Maple must first convert the expression to a series of commands of hardware floating-point numbers, and then convert the result of these back to Maple's format for numbers.

    Write a procedure that generates the coordinates of the grid in the form of an array. Since the procedure is to plot surfaces, the array is two-dimensional. The following procedure returns an array z of function values.

```
> evalgrid := proc( F, z, a, b, c, d, m, n )
>     local i, j, dx, dy;
>
>     dx := (b-a)/m; dy := (d-c)/n;
>     for i to m do
>        for j to n do
>           z[i, j] := F( a + (i-1)*dx, c + (j-1)*dy );
```

```
>         od;
>      od;
> end:
```

This `evalgrid` procedure is purely symbolic and does not handle error conditions.

```
> A := array(1..2, 1..2):
> evalgrid( f, 'A', 1, 2, 1, 2, 2, 2 ):
> eval(A);
```

$$\left[\begin{array}{cc} \dfrac{1}{\sin(1)} & \dfrac{1}{\sin(\frac{3}{2})} \\[2ex] \dfrac{1}{\sin(\frac{3}{2})} & \dfrac{1}{\sin(\frac{9}{4})} \end{array} \right]$$

```
> evalgrid( f, 'A', 0, Pi, 0, Pi, 15, 15 ):
Error, (in f) division by zero
```

Write a second procedure, `gridpoints`, which makes use of `evalgrid`. The procedure should accept a function, two ranges, and the number of grid points to generate in each dimension. Like the procedure `fnum` which Maple generated from your function f above, this routine should attempt to create the grid using hardware floating-point arithmetic. Only if this fails, should `gridpoints` resort to software floating-point arithmetic.

```
> gridpoints := proc( f, r1, r2, m, n )
>    local u, x, y, z, a, b, c, d;
>
>    # Domain information:
>    a := lhs(r1); b := rhs(r1);
>    c := lhs(r2); d := rhs(r2);
>
>    z := hfarray(1..m, 1..n);
>    if Digits <= evalhf(Digits) then
>        # Try to use hardware floats
>        # - notice the need for var in this case.
>        u := traperror( evalhf( evalgrid(f, var(z),
>            a, b, c, d, m, n) ) );
>        if lasterror = u then
>            # Use software floats, first converting f to
>            # a software float function.
>            z := array( 1..m, 1..n );
>            evalgrid( convert( f, numericproc ),
>                    z, a, b, c, d, m, n );
>        fi;
>    else
>        # Use software floats, first converting f to
>        # a software float function.
```

```
>          z := array( 1..m, 1..n );
>          evalgrid( convert(f, numericproc), z,
>              a, b, c, d, m, n );
>    fi;
>    eval(z);
> end:
```

The second argument to evalgrid must be the array (or hfarray) which receives the results; Maple must not convert it to a number before it calls evalhf. Indicate this special status to Maple using the special function var whenever you call evalgrid from within evalhf. Chapter 7 discusses numerical calculations in detail.

Test the procedures. Here gridpoints can use hardware floating-point arithmetic to calculate two of the numbers, but it must resort to software calculations in four cases where the function turns out to be undefined.

```
> gridpoints( (x,y) -> 1/sin(x*y) , 0..3, 0..3, 2, 3 );
```

$$[\textit{undefined}, \textit{undefined}, \textit{undefined}]$$

$$[\textit{undefined}, 1.00251130424672485,$$

$$7.08616739573718668]$$

In the following example, gridpoints can use hardware floating-point for all the calculations. Therefore, this calculation is faster, although the difference will not be apparent unless you try a much larger example.

```
> gridpoints( (x,y) -> sin(x*y) , 0..3, 0..3, 2, 3 );
```

$$[[0, 0, 0], [0, 0.997494986604054, 0.141120008059867]$$

$$]$$

If you ask for more digits than hardware floating-point arithmetic can provide, then gridpoints must always use software floating-point operations.

```
> Digits := 22:
> gridpoints( (x,y) -> sin(x*y) , 0..3, 0..3, 2, 3 );
```

$$[0, 0, 0]$$

$$[0, .9974949866040544309417,$$

$$.1411200080598672221007]$$

```
> Digits := 10:
```

The `gridpoints` procedure is remarkably similar to the `convert(...,` `gridpoints)` procedure which is part of the standard Maple library. The library command includes more checking of the arguments and, therefore, will likely suffice for many of your needs.

8.8 Animation

Maple has the ability to generate animations in either two or three dimensions. As with all of Maple's plotting facilities, such animations produce user-accessible data structures. Data structures of the following type represent animations.

```
PLOT( ANIMATE( ... ) )
```

or

```
PLOT3D( ANIMATE( ... ) )
```

Inside the `ANIMATE` function is a sequence of frames; each frame is a list of the same plotting objects that can appear in a single plotting structure. Every procedure that creates an animation builds such a sequence of frames. You can see an example by printing the output of such a procedure.

```
> lprint( plots[animate]( x*t, x=-1..1, t = 1..3,
>           numpoints=3, frames = 3 ) );

PLOT(ANIMATE([CURVES([[-1., -1.], [0, 0], [1.000000000
, 1.]],COLOUR(RGB,1.00000000,0,0))],[CURVES([[-1., -2.
], [0, 0], [1.000000000, 2.]],COLOUR(RGB,1.00000000,0,
0))],[CURVES([[-1., -3.], [0, 0], [1.000000000, 3.]],
COLOUR(RGB,1.00000000,0,0))]),AXESLABELS(x,''),VIEW(-1\
. .. 1.,DEFAULT))
```

The function `points` below is a parametrization of the curve $(x, y) = (1 + \cos(t\pi/180)^2, 1 + \cos(t\pi/180)\sin(t\pi/180))$.

```
> points := t -> evalf(
>           [ (1 + cos(t/180*Pi)) * cos(t/180*Pi ),
>             (1 + cos(t/180*Pi)) * sin(t/180*Pi ) ] ):
```

For example,

```
> points(2);
```

$$[1.998172852, .06977773357]$$

You can plot a sequence of points.

```
> PLOT( POINTS( seq( points(t), t=0..90 ) ) );
```

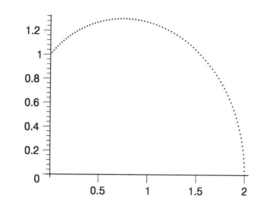

You can now make an animation. Make each frame consist of the polygon spanned by the origin, $(0, 0)$, and the sequence of points on the curve.

```
> frame := n -> [ POLYGONS([ [ 0, 0 ],
>                    seq( points(t), t = 0..60*n) ],
>                    COLOR(RGB, 1.0/n, 1.0/n, 1.0/n) ) ]:
```

The animation consists of six frames.

```
> PLOT( ANIMATE( seq( frame(n), n = 1..6 ) ) );
```

The display command from the plots package can show an animation in static form.

```
> with(plots):
```

```
> display( PLOT(ANIMATE(seq(frame(n), n = 1..6))) );
```

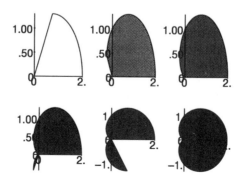

The `varyAspect` procedure below illustrates how a stellated surface varies with the aspect ratio. The procedure takes a graphical object as input and creates an animation in which each frame is a stellated version of the object with a different aspect ratio.

```
> with(plottools):
> varyAspect := proc( p )
>    local n, opts;
>    opts := convert( [ args[2..nargs] ], PLOT3Doptions );
>    PLOT3D( ANIMATE( seq( [ stellate( p, n/sqrt(2)) ],
>                          n=1..4 ) ),
>            op( opts ));
> end:
```

Try the procedure on a dodecahedron.

```
> varyAspect( dodecahedron(), scaling=constrained );
```

Here is the static version.

```
> display( varyAspect( dodecahedron(),
>                        scaling=constrained ) );
```

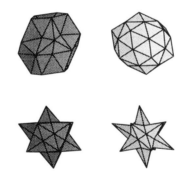

The Maple library provides three methods for creating animations: the `animate` and `animate3d` commands in the `plots` package, or the `display` command with the `insequence = true` option set. For example, you can show how a Fourier series approximates a function, f, on an interval $[a, b]$ by visualizing the function and successive approximations as the number of terms increase with each frame. You can derive the nth partial sum of the Fourier series using $f_n(x) = c_0/2 + \sum_{k=1}^{n} c_k \cos(\frac{2\pi}{b-a}kx) + s_k \sin(\frac{2\pi}{b-a}kx)$, where

$$c_k = \frac{2}{b-a} \int_a^b f(x) \cos\left(\frac{2\pi}{b-a}kx \right) dx$$

and

$$s_k = \frac{2}{b-a} \int_a^b f(x) \sin\left(\frac{2\pi}{b-a}kx\right) dx.$$

The `fourierPicture` procedure below first calculates and plots the kth Fourier approximation for k up to n. Then `fourierPicture` generates an animation of these plots, and finally it adds a plot of the function itself as a backdrop.

```
> fourierPicture :=
> proc( func, xrange::name=range, n::posint)
>    local x, a, b, l, k, j, p, q, partsum;
>
>    a := lhs( rhs(xrange) );
>    b := rhs( rhs(xrange) );
>    l := b - a;
>    x := 2 * Pi * lhs(xrange) / l;
>
>    partsum := 1/l * evalf( Int( func, xrange) );
>    for k from 1 to n do
>       # Generate the terms of the Fourier series of func.
>       partsum := partsum
>          + 2/l * evalf( Int(func*sin(k*x), xrange) )
>                * sin(k*x)
>          + 2/l * evalf( Int(func*cos(k*x), xrange) )
>                * cos(k*x);
>       # Plot k-th Fourier approximation.
>       q[k] := plot( partsum, xrange, color=blue,
>                         args[4..nargs] );
>    od;
>    # Generate sequence of frames.
>    q := plots[display]( [ seq( q[k], k=1..n ) ],
>                          insequence=true );
>    # Add the function plot, p, to each frame.
>    p := plot( func, xrange, color = red, args[4..nargs] );
>    plots[display]( [ q, p ] );
> end:
```

You can now use `fourierPicture` to see, for example, the first six Fourier approximations of e^x.

```
> fourierPicture( exp(x), x=0..10, 6 );
```

This is the static version.

```
> display( fourierPicture( exp(x), x=0..10, 6 ) );
```

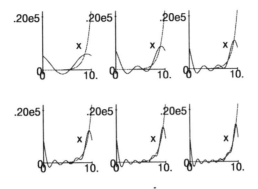

Below are the first six Fourier approximations of x -> signum(x-1). The signum function is discontinuous, so the discont=true option is called for.

```
> fourierPicture( 2*signum(x-1), x=-2..3, 6,
>                 discont=true );
```

Again, these pages require a static version.

```
> display( fourierPicture( 2*signum(x-1), x=-2..3, 6,
>                 discont=true ) );
```

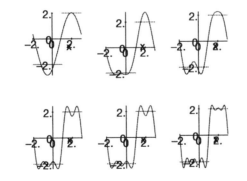

You can also create similar animations with other series approximations, such as Taylor, Padé, and Chebyshev–Padé, with the generalized series structures that Maple uses.

Animation sequences exist in both two and three dimensions. The procedure below ties a trefoil knot using the tubeplot function in the plots package.

```
> TieKnot := proc( n:: posint )
```

```
>    local i, t, curve, picts;
>    curve := [ -10*cos(t) - 2*cos(5*t) + 15*sin(2*t),
>                 -15*cos(2*t) + 10*sin(t) - 2*sin(5*t),
>                 10*cos(3*t) ]:
>    picts := [ seq( plots[tubeplot]( curve,
>                                 t=0..2*Pi*i/n, radius=3),
>                    i=1..n ) ];
>    plots[display]( picts, insequence=true, style=patch);
> end:
```

You can tie the knot in, say, six stages.

```
> TieKnot(6);
```

Here is the static version.

```
> display( TieKnot(6) );
```

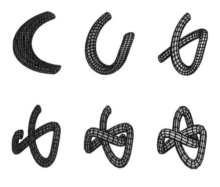

You can combine the graphical objects from the plottools package with the display in-sequence option to animate physical objects in motion. The springPlot procedure below creates a spring from a three-dimensional plot of a helix. springPlot also creates a box and a copy of this box and moves one of the boxes to various locations depending on a value of u. For every u, you can locate these boxes above and below the spring. Finally springPlot makes a sphere and translates it to locations above the top of the top box with the height again varying with a parameter. Finally, it produces the entire animation by organizing a sequence of positions and showing them in sequence using display.

```
> springPlot := proc( n )
>    local u, curve, springs, box, tops, bottoms,
>          helix, ball, balls;
>    curve := (u,v) -> spacecurve(
>          [cos(t), sin(t), 8*sin(u/v*Pi)*t/200],
>          t=0..20*Pi,
```

```
>              color=black, numpoints=200, thickness=3 ):
>    springs := display( [ seq(curve(u,n), u=1..n) ],
>                        insequence=true ):
>    box := cuboid( [-1,-1,0], [1,1,1], color=red ):
>    ball := sphere( [0,0,2], grid=[15, 15], color=blue ):
>    tops :=  display( [ seq(
>      translate( box, 0, 0, sin(u/n*Pi)*4*Pi/5 ),
>      u=1..n ) ], insequence=true ):
>    bottoms := display( [ seq( translate(box, 0, 0, -1),
>      u=1..n ) ], insequence=true ):
>    balls := display( [ seq( translate( ball, 0, 0,
>         4*sin( (u-1)/(n-1)*Pi ) + 8*sin(u/n*Pi)*Pi/10 ),
>         u=1..n ) ],   insequence=true ):
>    display( springs, tops, bottoms, balls,
>        style=patch, orientation=[45,76],
>        scaling=constrained );
> end:
```

The code above uses the short names of the commands from the plots
and plottools packages in order to improve readability. You must either use long names or remember to load these two packages before using
springPlot.

```
> with(plots): with(plottools):
> springPlot(6);
> display( springPlot(6) );
```

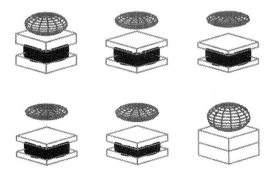

Programming with the plottools *Package* on page 285 describes how
the commands in the plottools package can help you with graphics procedures.

8.9 Programming with Color

As well as coloring each type of object in the plot data structures, you can also add colors to plotting routines. The color option allows you to specify colors in the form of a solid color, by name, by RGB or HUE values, or via a color function in the form of a Maple formula or function. Try each of the following commands for yourself.

```
> plot3d( sin(x*y), x=-3..3, y=-3..3, color=red );
> plot3d( sin(x*y), x=-3..3, y=-3..3,
>    color=COLOUR(RGB, 0.3, 0.42, 0.1) );

> p := (x,y) -> sin(x*y):
> q := (x,y) -> if x < y then 1 else x - y fi:

> plot3d( p, -3..3, -3..3, color=q );
```

Although usually less convenient, you may also specify the color attributes at the lower level of graphics primitives. At the lowest level, you can accomplish a coloring of a graphical object by including a COLOUR function as one of the options inside the object.

```
> PLOT( POLYGONS( [ [0,0], [1,0], [1,1] ],
>                 [ [1,0], [1,1], [2,1], [2,0] ],
>                 COLOUR(RGB, 1/2, 1/3, 1/4 ) ) );
```

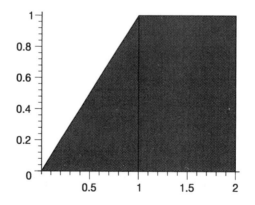

You can use different colors for each polygon via either

```
PLOT( POLYGONS( P1, ... , Pn ,
COLOUR(RGB, p1, ..., pn)) )
```

or

```
PLOT( POLYGONS( P1, COLOUR(RGB, p1) ), ... ,
POLYGONS( Pn, COLOUR(RGB, pn)) )
```

Thus, the following two PLOT structures represent the same picture of a red and a green triangle.

```
> PLOT( POLYGONS( [ [0,0], [1,1], [2,0] ],
>                  COLOUR( RGB, 1, 0, 0 ) ),
>        POLYGONS( [ [0,0], [1,1], [0,1] ],
>                  COLOUR( RGB, 0, 1, 0 ) ) );

> PLOT( POLYGONS( [ [0,0], [1,1], [2,0] ],
>                  [ [0,0], [1,1], [0,1] ],
>                  COLOUR( RGB, 1, 0, 0, 0, 1, 0 ) ) );
```

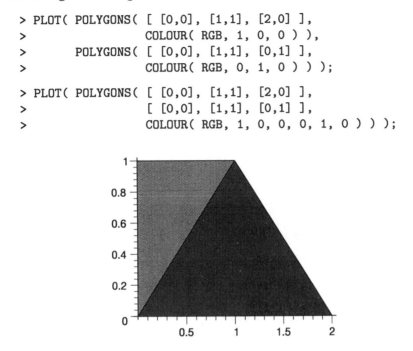

The three RGB values must be numbers between 0 and 1.

Generating Color Tables

The following procedure generates an $m \times n$ color table of RGB values. Specifically, colormap returns a sequence of two elements: a POLYGONS structure and a TITLE.

```
> colormap := proc(m, n, B)
>    local i, j, points, colors, flatten;
>    # points = sequence of corners for rectangles
>    points :=  seq( seq( evalf(
>            [ [i/m, j/n], [(i+1)/m, j/n],
>              [(i+1)/m, (j+1)/n], [i/m, (j+1)/n] ]
>                 ), i=0..m-1 ), j=0..n-1 ):
>    # colors = listlist of RGB color values
>    colors :=  [seq( seq( [i/(m-1), j/(n-1), B],
>                    i=0..m-1 ), j=0..n-1 )] ;
>    # flatten turns the colors listlist into a sequence
>    flatten := a -> op( map(op, a) );
>    POLYGONS( points,
>              COLOUR(RGB, flatten(colors) ) ),
```

```
>    TITLE( cat( "Blue=", convert(B, string) ) );
> end:
```

Here is a 10 × 10 table of colors; the blue component is 0.

```
> PLOT( colormap(10, 10, 0) );
```

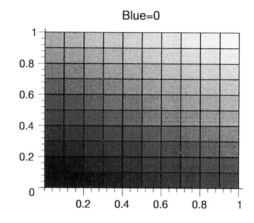

You can use animation to vary the blue component as well. The `colormaps` procedure below uses animation to generate an $m \times n \times f$ color table.

```
> colormaps := proc(m, n, f)
>    local t;
>    PLOT( ANIMATE( seq( [ colormap(m, n, t/(f-1)) ],
>                        t=0..f-1 ) ),
>          AXESLABELS("Red", "Green") );
> end:
```

The following gives you a 10 × 10 × 10 color table.

```
> colormaps(10, 10, 10);
```

You can visualize the color scale for HUE coloring as follows.

```
> points := evalf( seq( [ [i/50, 0], [i/50, 1],
>                         [(i+1)/50, 1], [(i+1)/50, 0] ],
>                         i=0..49)):

> PLOT( POLYGONS(points, COLOUR(HUE, seq(i/50, i=0..49)) ),
>       AXESTICKS(DEFAULT, 0), STYLE(PATCHNOGRID) );
```

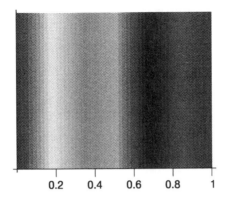

The AXESTICKS(DEFAULT, 0) specification eliminates the axes labeling along the vertical axes but leaves the default labeling along the horizontal axis.

You can easily see how to create a colormapHue procedure which creates the color scale for any color function based on HUE coloring.

```
> colormapHue := proc(F, n)
>    local i, points;
>    points := seq( evalf( [ [i/n, 0], [i/n, 1],
>                            [(i+1)/n, 1], [(i+1)/n, 0] ]
>                    ), i=0..n-1 ):
>    PLOT( POLYGONS( points,
>            COLOUR(HUE, seq( evalf(F(i/n)), i=0.. n-1) )),
>          AXESTICKS(DEFAULT, 0), STYLE(PATCHNOGRID) );
> end:
```

The basis of this color scale is $y(x) = \sin(\pi x)/3$ for $0 \le x \le 40$.

```
> colormapHue( x -> sin(Pi*x)/3, 40);
```

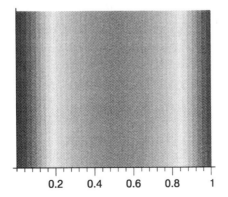

Visualizing the gray scale coloring is a simple matter of using an arbitrary procedure, F, since gray levels are simply those levels that have equal parts of red, green, and blue.

```
> colormapGraylevel := proc(F, n)
>    local i, flatten, points, grays;
>    points := seq( evalf([ [i/n, 0], [i/n, 1],
>                           [(i+1)/n, 1], [(i+1)/n, 0] ]),
>                 i=0..n-1):
>    flatten := a -> op( map(op, a) );
>    grays := COLOUR(RGB, flatten(
>          [ seq( evalf([ F(i/n), F(i/n), F(i/n) ]),
>                i=1.. n)]));
>    PLOT( POLYGONS(points, grays),
>          AXESTICKS(DEFAULT, 0) );
> end:
```

The identity function, $x \mapsto x$, yields the basic gray scale.

```
> colormapGraylevel( x->x, 20);
```

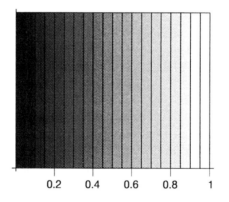

Adding Color Information to Plots

You can add color information to an existing plot data structure. The procedure addCurvecolor colors each curve in a CURVES function via the scaled y coordinates.

```
> addCurvecolor := proc(curve)
>    local i, j, N, n , M, m, curves, curveopts, p, q;
>
>    # Get existing point information.
>    curves := select( type, [ op(curve) ],
>                    list(list(numeric)) );
>    # Get all options but color options.
```

```
>     curveopts := remove( type, [ op(curve) ],
>                         { list(list(numeric)),
>                           specfunc(anything, COLOR),
>                           specfunc(anything, COLOUR) } );
>
>     # Determine the scaling.
>     # M and m are the max and min of the y-coords.
>     n := nops( curves );
>     N := map( nops, curves );
>     M := [ seq( max( seq( curves[j][i][2],
>             i=1..N[j] ) ), j=1..n ) ];
>     m := [ seq( min( seq( curves[j][i][2],
>             i=1..N[j] ) ), j=1..n ) ];
>     # Build new curves adding HUE color.
>     seq( CURVES( seq( [curves[j][i], curves[j][i+1]],
>                      i=1..N[j]-1 ),
>               COLOUR(HUE, seq((curves[j][i][2]
>                               - m[j])/(M[j] - m[j]),
>                          i=1..N[j]-1)),
>               op(curveopts) ), j=1..n );
> end:
```

For example

```
> c := CURVES( [ [0,0], [1,1], [2,2], [3,3] ],
>              [ [2,0], [2,1], [3,1] ] );
```

$$c := \text{CURVES}([[0, 0], [1, 1], [2, 2], [3, 3]],$$

$$[[2, 0], [2, 1], [3, 1]])$$

```
> addCurvecolor( c );
```

$$\text{CURVES}([[0, 0], [1, 1]], [[1, 1], [2, 2]], [[2, 2], [3, 3]],$$

$$\text{COLOUR}\left(HUE, 0, \frac{1}{3}, \frac{2}{3}\right)), \text{CURVES}([[2, 0], [2, 1]],$$

$$[[2, 1], [3, 1]], \text{COLOUR}(HUE, 0, 1))$$

You can then map such a procedure over all CURVES structures of an existing plot structure to provide the desired coloring for each curve.

```
> addcolor := proc( aplot )
>     local recolor;
>     recolor := x -> if op(0,x)=CURVES then
>                       addCurvecolor(x)
>                     else x fi;
```

```
>    map( recolor, aplot );
> end:
```

Try addcolor on a plot of $\sin(x) + \cos(x)$.

```
> p := plot( sin(x) + cos(x), x=0..2*Pi,
>               linestyle=2, thickness=3 ):
> addcolor( p );
```

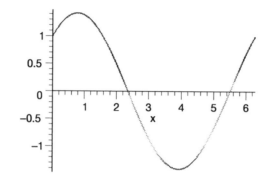

If you add color to two curves simultaneously, the two colorings are independent.

```
> q := plot( cos(2*x) + sin(x), x=0..2*Pi ):
> addcolor( plots[display](p, q) );
```

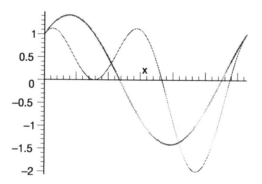

The addcolor procedure also works on three-dimensional space curves.

```
> spc := plots[spacecurve]( [ cos(t), sin(t), t ],
>                  t=0..8*Pi, thickness=2, color=black ):
```

```
> addcolor( spc );
```

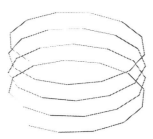

You can easily alter the coloring of an existing plot using coloring functions. Such coloring functions should either be of the form $C_{Hue}: R^2 \rightarrow$ [0, 1] (for Hue coloring) or of the form $C_{RGB}: R^2 \rightarrow [0, 1] \times [0, 1] \times [0, 1]$.
The example above uses the color function $C_{Hue}(x, y) = y/\max(y_i)$.

Creating A Chess Board Plot

The final example of programming with color shows how to make a chess board type grid with red and white squares in a three-dimensional plot. You do not simply assign a coloring function as an argument to plot3d. A coloring function, in such a case, provides colors for vertices of a grid, which does not yield color patches. You must first convert the grid or mesh into polygonal form. The rest of the procedure assigns either a red or white color to a polygon, depending on which grid area it represents.

```
> chessplot3d := proc(f, r1, r2)
>     local m, n, i, j, plotgrid, p, opts, coloring, size;
>
>     # obtain grid size
>     # and generate the plotting data structure
>     if hasoption( [ args[4..nargs] ], grid, size) then
>         m := size[1];
>         n := size[2];
>     else  # defaults
>         m := 25;
>         n := 25;
>     fi;
>
>     p := plot3d( f, r1, r2, args[4..nargs] );
>
>     # convert grid data (first operand of p)
```

```
>      # into polygon data
>      plotgrid := op( convert( op(1, p), POLYGONS ) );
>      # make coloring function - alternating red and white
>      coloring := (i, j) -> if modp(i-j, 2)=0 then
>                                convert(red, colorRGB)
>                          else
>                                convert(white, colorRGB)
>                          fi;
>      # op(2..-1, p) is all the operands of p but the first
>      PLOT3D( seq( seq( POLYGONS( plotgrid[j + (i-1)*(n-1)],
>                                coloring(i, j) ),
>                 i=1..m-1 ), j=1..n-1 ),
>             op(2..-1, p) );
> end:
```

Here is a chess board plot of $\sin(x)\sin(y)$.

```
> chessplot3d( sin(x)*sin(y), x=-Pi..Pi, y=-Pi..Pi,
>              style=patch, axes=frame );
```

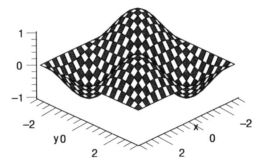

Note that chessplot3d works when the plotting structure from plot3d is either a GRID or MESH output type. The latter is the type of output that comes from parametric surfaces or from surfaces that use alternate coordinate systems.

```
> chessplot3d( (4/3)^x*sin(y), x=-1..2*Pi, y=0..Pi,
>              coords=spherical, style=patch,
>              lightmodel=light4 );
```

8.10 Conclusion

In this chapter, you have seen how you can make graphics procedures based on the commands `plot` and `plot3d`, as well as the commands found in the `plots` and `plottools` packages. However, for ultimate control, you must create PLOT and PLOT3D data structures directly; these are the primitive specifications of all Maple plots. Inside the PLOT and PLOT3D data structures you can specify points, curves, and polygons, as well as grids of values and meshes of points. You have also seen how to handle plot options, create numerical plotting procedures, work with grids and meshes, manipulate plots and animations, and apply non-standard coloring to your graphics.

Input and Output

Although Maple is primarily a system and language for performing mathematical manipulations, many situations arise where such manipulations require the use of data from outside of Maple, or the production of data in a form suitable for use by other applications. You may also need Maple programs to request input directly from the user and/or present output directly to the user. To meet these needs, Maple provides a comprehensive collection of input and output (I/O) commands. The *Maple I/O library* is the term which refers to these commands as a group.

9.1   A Tutorial Example

This section illustrates some of the ways you can use the Maple I/O library in your work. Specifically, the examples show how to write a table of numerical data to a file, and how to read such a table from a file. The examples refer to the following data set, given in the form of a list of lists and assumed to represent a list of (x, y) pairs, where each x is an integer and each y is a real number.

```
> A := [[0, 0],
>       [1, .8427007929],
>       [2, .9953222650],
>       [3, .9999779095],
>       [4, .9999999846],
>       [5, 1.000000000]]:
```

In a real application, this list would have been generated by a Maple command you executed or by a Maple procedure you wrote. In this example, the list was simply typed in as you see it above.

If you want to use some other program (like a presentation graphics program, or perhaps a custom C program) to process data that Maple has generated, then you often need to save the data to a file in a format that the other program recognizes. Using the I/O library, you will find it easy to write such data to a file.

```
> for xy in A do fprintf("myfile", "%d %e\n", xy[1], xy[2]) od:
> fclose("myfile");
```

If you print the file myfile, or view it with a text editor, it looks like this:

```
0 0.000000e-01
1 8.427008e-01
2 9.953223e-01
3 9.999779e-01
4 1.000000e+00
5 1.000000e+00
```

The fprintf command wrote each pair of numbers to the file. This command takes two or more arguments, the first of which specifies the file that Maple is to write, and the second of which specifies the format for the data items. The remaining arguments are the actual data items that Maple is to write.

In the example above, the file name is myfile. The first time a given file name appears as an argument to fprintf (or any of the other output commands described later), the command creates the file if it does not already exist, and prepares (opens) it for writing. If the file exists, the new version overwrites the old one. You can override this behavior (for example, if you want to append to an already existing file) using the fopen command, described later.

The format string, "%d %e\n", specifies that Maple should write the first data item as a decimal integer (%d), and the second data item in FORTRAN-like scientific notation (%e). A single space should separate the first and second data items, and a line break (\n) should follow the second data item (to write each pair of numbers on a new line). By default, as in our example, Maple rounds floating-point numbers to six significant digits for output. You can specify more or fewer digits using options to the %e format. The section on fprintf describes these options in more detail.

When you are finished writing to a file, you must close it. Until you close a file, the data may or may not actually be in the file, because output is buffered under most operating systems. The fclose command closes a file. If you forget to close a file, Maple automatically closes it when you exit.

For a simple case like the one presented here, writing the data to a file using the `writedata` command is easier.

```
> writedata("myfile2", A, [integer,float]);
```

The `writedata` command performs all the operations of opening the file, writing the data in the specified format, an integer and a floating-point number, and closing the file. However, `writedata` does not provide the precise formatting control that you may need in some cases. For this, use `fprintf` directly.

In some applications, you may want to read data from a file. For example, some data acquisition software may supply data that you may want to analyze. Reading data from a file is almost as easy as writing it.

```
> A := [];
```

$$A := []$$

```
> do
>     xy := fscanf("myfile2", "%d %e");
>     if xy = 0 then break fi;
>     A := [op(A),xy];
> od;
```

$$xy := [0, 0]$$
$$A := [[0, 0]]$$
$$xy := [1, .842701]$$
$$A := [[0, 0], [1, .842701]]$$
$$xy := [2, .995322]$$
$$A := [[0, 0], [1, .842701], [2, .995322]]$$
$$xy := [3, .999978]$$
$$A := [[0, 0], [1, .842701], [2, .995322], [3, .999978]]$$
$$xy := [4, 1.]$$

$$A :=$$
$$[[0, 0], [1, .842701], [2, .995322], [3, .999978], [4, 1.]]$$
$$xy := [5, 1.]$$

$$A := [[0, 0], [1, .842701], [2, .995322], [3, .999978],$$
$$[4, 1.], [5, 1.]]$$

$$xy := []$$

$$A := [[0, 0], [1, .842701], [2, .995322], [3, .999978],$$

$$[4, 1.], [5, 1.], []]$$

$$xy := 0$$

```
> fclose("myfile2");
```

This example starts by initializing A to be the empty list. Upon entering the loop, Maple reads a pair of numbers at a time from the file.

The fscanf command reads characters from a specified file, and parses them according to the specified format (in this case, "%d %e", indicating a decimal integer and a real number). It either returns a list of the resulting values or the integer 0 to indicate that it has reached the end of the file. The first time you call fscanf with a given file name, Maple prepares (opens) the file for reading. If it does not exist, Maple generates an error.

The second line of the loop checks if fscanf returned 0 to indicate the end of the file, and breaks out of the loop if it has. Otherwise, Maple appends the pair of numbers to the list of pairs in A. (The syntax A := [op(A),xy] tells Maple to assign to A a list consisting of the existing elements of A, and the new element xy.)

As when you wrote to a file, you can read from a file more easily using the readdata command.

```
> A := readdata("myfile2", [integer,float]);
```

$$A := [[0, 0], [1, .842701], [2, .995322], [3, .999978],$$

$$[4, 1.], [5, 1.]]$$

The readdata command performs all the operations of opening the file, reading the data and parsing the specified format (an integer and a floating-point number), and closing the file. However, readdata does not provide the precise parsing control that you may need in some cases. For this, use fscanf directly.

These examples illustrate some of the basic concepts of Maple's I/O library, and you can do a great deal using only the information presented in this section. However, to make more effective and efficient use of the I/O library, an understanding of a few more concepts and commands is useful. The remainder of this chapter describes the concepts of file types, modes, descriptors, and names, and presents a variety of commands for performing both formatted and unformatted file I/O.

9.2   File Types and Modes

Most of the Maple I/O library commands operate on files. This chapter uses the term *file* to denote not just files on a disk, but also Maple's user interface. In most cases, you cannot distinguish between the two from the point of view of the I/O commands. Almost any operation that you can perform on a real file you can perform on the user interface, if appropriate.

Buffered Files versus Unbuffered Files

The Maple I/O library can deal with two different kinds of files: buffered (STREAM) and unbuffered (RAW). No difference exists in how Maple uses them, but buffered files are usually faster. In buffered files, Maple collects characters in a buffer and writes them to a file all at once when the buffer is full or the file is closed. Raw files are useful when you wish to explicitly take advantage of knowledge about the underlying operating system, such as the block size on the disk. For general use, you should use buffered files, and they are used by default by most of the I/O library commands.

Commands that provide information about I/O status use the identifiers STREAM and RAW to indicate buffered and unbuffered files, respectively.

Text Files versus Binary Files

Many operating systems, including DOS/Windows, the Macintosh operating system, and VMS, distinguish between files containing sequences of characters (*text files*) and files containing sequences of bytes (*binary files*). The distinction lies primarily in the treatment of the new line character. Other distinctions may exist on some platforms, but they are not visible when using the Maple I/O library.

Within Maple, the new line character, which represents the concept of ending one line and beginning a new one, is a single character (although you can type it as the two characters "\n" within Maple strings). The internal representation of this character is the byte whose value is 10, the ASCII line feed character. Many operating systems, however, represent the concept of new line within a file using a different character, or a sequence of two characters. For example, DOS/Windows and VMS represent a new line with two consecutive bytes whose values are 13 and 10 (carriage return and line feed). The Macintosh represents a new line with the single byte with value 13 (carriage return).

The Maple I/O library can deal with files as either text files or binary files. When Maple writes to a text file, any new line characters that it writes to the file are translated into the appropriate character or character sequence that the underlying operating system uses. When Maple reads this character or character sequence from a file, it translates back into the single new line character. When Maple writes to a binary file, no translation takes place; it reads new line characters and writes them as the single byte with value 10.

When running Maple under the UNIX operating system or one of its many variants, Maple makes no distinction between text and binary files. It treats both in the same way, and no translation takes place.

Commands which can specify or query whether a file is a text file or a binary file use the identifiers TEXT and BINARY, respectively.

Read Mode versus Write Mode

At any given time, a file may be open either for reading or for writing. You cannot write to a file that is open only for reading; but, you can write to and read from a file that is open for writing. If you attempt, using the Maple I/O library, to write to a file which is open only for reading, Maple closes and reopens the file for writing. If the user does not have the necessary permissions to write to the file (if the file is read-only, or resides on a read-only file system), errors occur at that point.

Commands where you can specify or query whether a file is open for reading or writing use the identifiers READ and WRITE, respectively.

The `default` and `terminal` Files

The Maple I/O library treats the Maple user interface as a file. The identifiers `default` and `terminal` refer to this file. The `default` identifier refers to the current input stream, the one from which Maple reads and processes commands. The `terminal` identifier refers to the top level input stream, the one which was the current input stream when you first started Maple.

When Maple is run interactively, `default` and `terminal` are equivalent. Only when reading commands from a source file using the `read` statement does a distinction arise. In that case, `default` refers to the file being read; whereas, `terminal` refers to the session. Under UNIX, if input is redirected from a file or pipe, `terminal` refers to that file or pipe.

Note that only the *symbols* `default` and `terminal` are special; the *strings* "default" and mname"terminal" simply refer to files with those names.

9.3 File Descriptors versus File Names

The commands of the Maple I/O library refer to files in one of two ways: by name or by descriptor.

Referring to a file by name is the simpler of the two methods. The first time Maple performs an operation on the file, it opens the file, either in READ mode or in WRITE mode and as a TEXT file or a BINARY file, as appropriate to the operation that it is performing. The primary advantage of referring to files by name is simplicity. However, you will experience a slight performance penalty for using this method, especially if performing many small operations on a file (such as writing individual characters).

Referring to a file by descriptor is only slightly more complex and is a familiar concept to those who have programmed in more traditional environments. A descriptor simply identifies a file after you have opened it. Use the name of the file once to open it and create a descriptor. When you subsequently manipulate the file, use the descriptor instead of the file name. An example in *Opening and Closing Files* on page 336 illustrates the use of a file descriptor.

The advantages of the descriptor method include more flexibility when opening the file (you can specify whether the file is TEXT or BINARY, and if Maple should open the file in READ mode or in WRITE mode), improved performance when performing many operations on a file, and the ability to work with unbuffered files. The disadvantage is a slight increase in the amount of programming that you must do.

Which approach is best depends on the task at hand. You can perform simple file I/O tasks most easily using names; whereas, more complex tasks can benefit from the use of descriptors.

In subsequent sections, the term *file/dentifier* refers to either a file name or a file descriptor.

9.4 File Manipulation Commands

Opening and Closing Files

Before you can read from or write to a file, you must open it. When referring to files by name, this happens automatically with the first attempt at any operation on the file. When you use descriptors, however, you must explicitly open the file first in order to create the descriptor.

The two commands for opening files are fopen and open. The fopen command opens buffered (STREAM) files; whereas, the open command opens unbuffered (RAW) files.

Use the fopen command as follows.

```
fopen( fileName, accessMode, fileType )
```

The *fileName* specifies the name of the file to open. This name is specified as a string, and follows the conventions that the underlying operating system uses. The *accessMode* must be one of READ, WRITE, or APPEND, indicating whether you should initially open the file for reading, writing, or appending. The optional *fileType* is either TEXT or BINARY.

If you try to open the file for reading and it does not exist, fopen generates an error (which can be trapped using traperror).

If you try to open the file for writing and it does not exist, Maple first creates it. If it does exist and you specify WRITE, Maple truncates the file to zero length; if you specified APPEND, subsequent calls to commands that write to the file append to it.

Call the open command as follows.

```
open( fileName, accessMode )
```

The arguments to open are the same as those to fopen, except that you cannot specify a *fileType* (TEXT or BINARY). Maple opens an unbuffered file with type BINARY.

Both fopen and open return a file descriptor. Use this descriptor to refer to the file for subsequent operations. You can still use the file's name, if you desire.

When you have finished with a file, you should tell Maple to close it. This ensures that Maple actually writes all information to the disk. It also frees up resources of the underlying operating system, which often imposes a limit on the number of files that you can open simultaneously.

Close files using the fclose or close commands. These two commands are equivalent, and you can call them as follows.

```
fclose( fileIdentifier )
close( fileIdentifier )
```

The *fileIdentifier* is the name or descriptor of the file you wish to close. Once you close a file, any descriptors referring to the file are no longer valid.

```
> f := fopen("testFile.txt",WRITE);
```

$$f := 0$$

```
> writeline(f,"This is a test");
```

15

```
> fclose(f);
```

```
> writeline(f,"This is another test");

Error, (in fprintf) file descriptor not in use
```

When you exit Maple or issue a `restart` command, Maple automatically closes any open files, whether you opened them explicitly using `fopen` or `open`, or implicitly through a file I/O command.

Position Determination and Adjustment

Associated with each open file is the concept of its current position. This is the location within the file to which a subsequent write takes place, or from which a subsequent read takes place. Any reading or writing operation advances the position by the number of bytes read or written.

You can determine the current position within a file using the `filepos` command. Use this command in the following manner.

> filepos(*fileIdentifier*, *position*)

The *fileIdentifier* is the name or descriptor of the file whose position you wish to determine or adjust. If you give a file name, and that file is not yet open, Maple opens it in READ mode with type BINARY.

The *position* is optional. If you do not specify the *position*, Maple returns the current position. If you supply the *position*, Maple sets the current position to your specifications and returns the resulting position. In that case, the returned position is the same as the specified *position* unless the file is shorter than the specified *position*, in which case the returned position is that of the end of the file (that is, its length). You can specify the *position* either as an integer, or as the name `infinity`, which specifies the end of the file.

The following command returns the length of the file `myfile.txt`.

```
> filepos("myfile.txt", infinity);
```

$$36$$

Detecting the End of a File

The `feof` command determines whether you have reached the end of a file. Only use the `feof` command on files that you have opened as STREAMs, either implicitly or explicitly via `fopen`. Call `feof` in the following manner.

> feof(*fileIdentifier*)

The *fileIdentifier* is the name or descriptor of the file that you wish to query. If you give a file name, and that file is not yet open, Maple opens it in READ mode with type BINARY.

The feof command returns true if and only if you have reached the end of the file during the most recent readline, readbytes, or fscanf operation. Otherwise, feof returns false. This means that if 20 bytes remain in a file and you use readbytes to read these 20 bytes, then feof still returns false. You only encounter the end-of-file when you attempt another read.

Determining File Status

The iostatus command returns detailed information about all the files currently in use. Call the iostatus command with the following syntax.

```
iostatus()
```

The iostatus command returns a list. The list contains the following elements:

iostatus()[1]   The number of files that the Maple I/O library is currently using.

iostatus()[2]   The number of active nested read commands (when read reads a file which itself contains a read statement).

iostatus()[3]   The upper bound on iostatus()[1] + iostatus()[2] that the underlying operating system imposes.

iostatus()[*n*]   for n > 3. A list giving information about a file currently in use by the Maple I/O library.

When $n > 3$, the lists that iostatus()[*n*] return each contain the following elements:

iostatus()[*n*][1]   The file descriptor which fopen or open returned.

iostatus()[*n*][2]   The name of the file.

iostatus()[*n*][3]   The kind of file (STREAM, RAW, or DIRECT).

iostatus()[*n*][4]   The file pointer or file descriptor that the underlying operating system uses. The pointer is in the form FP=*integer* or FD=*integer*.

iostatus()[*n*][5]   The file mode (READ or WRITE).

iostatus()[*n*][6]   The file type (TEXT or BINARY).

Removing Files

Many files are solely for temporary use. Often, you no longer need such files when you complete your Maple session and thus, you should remove them. Use the `fremove` command to do this.

```
fremove( fileIdentifier )
```

The *fileIdentifier* is the name or descriptor of the file you wish to remove. If the file is currently open, Maple closes it before removing it. If the file does not exist, Maple generates an error.

To remove a file without knowing whether it exists or not, use `traperror` to trap the error that `fremove` might create.

```
> traperror(fremove("myfile.txt")):
```

9.5 Input Commands

Reading Text Lines from a File

The `readline` command reads a single line of text from a file. Characters are read up to and including a new line. The `readline` command then discards the new line character, and returns the line of characters as a Maple string. If `readline` cannot read a whole line from the file, then it returns 0 instead of a string.

Call the `readline` command using the following syntax.

```
readline( fileIdentifier )
```

The *fileIdentifier* is the name or descriptor, of the file that you wish to read. For compatibility with earlier versions of Maple, you can omit the *fileIdentifier*, in which case Maple uses `default`. Thus `readline()` and `readline(default)` are equivalent.

If you use `-1` as the *fileIdentifier*, Maple also takes input from the `default` stream, except that Maple's command line preprocessor runs on all input lines. This means that lines beginning with "!" pass to the operating system instead of returning through `readline`, and that lines beginning with "?" translate to calls to the `help` command.

If you call `readline` with a file name, and that file is not yet open, Maple opens it in READ mode as a TEXT file. If `readline` returns 0 (indicating the end of the file) when called with a file name, it automatically closes the file.

The following example defines a Maple procedure which reads a text file and displays it on the `default` output stream.

```
> ShowFile := proc( fileName::string )
```

```
>      local line;
>      do
>          line := readline(fileName);
>          if line = 0 then break fi;
>          printf("%s\n",line);
>      od;
> end:
```

Reading Arbitrary Bytes from a File

The readbytes command reads one or more individual characters or bytes from a file, returning either a string or a list of integers. If there are no more characters remaining in the file when you call readbytes, the command returns 0, indicating that you have reached the end of the file.

Use the following syntax to call the readbytes command.

```
readbytes( fileIdentifier, length, TEXT )
```

The *fileIdentifier* is the name or descriptor of the file that Maple is to read. The *length*, which you may omit, specifies how many bytes Maple needs to read. If you omit *length*, Maple reads one byte. The optional parameter TEXT indicates that the result is to be returned as a string rather than a list of integers.

You can specify the *length* as infinity, in which case Maple reads the remainder of the file.

If you specify TEXT when a byte with value 0 resides among the bytes being read, the resulting string contains only those characters preceding the 0 byte.

If you call readbytes with a file name, and that file is not yet open, Maple opens it in READ mode. If you specify TEXT, Maple opens it as a TEXT file; otherwise, Maple opens it as a BINARY file. If readbytes returns 0 (indicating the end of the file) when you call it with a file name, it automatically closes the file.

The following example defines a Maple procedure which reads an entire file, using readbytes, and copies it to a new file.

```
> CopyFile := proc( sourceFile::string, destFile::string )
>     writebytes(destFile, readbytes(sourceFile, infinity))
> end:
```

Formatted Input

The fscanf and scanf commands read from a file, parsing numbers and substrings according to a specified format. The commands return a list of these parsed objects. If no more characters remain in the file when you call `fscanf` or `scanf`, they return 0 instead of a list, indicating that it has reached the end of the file.

Call the `fscanf` and `scanf` commands as follows.

> `fscanf( fileIdentifier, format )`
> `scanf( format )`

The *fileIdentifier* is the name or descriptor of the file you wish to read. A call to `scanf` is equivalent to a call to `fscanf` with `default` as the *fileIdentifier*.

If you call `fscanf` with a file name, and that file is not yet open, Maple opens it in READ mode as a TEXT file. If `fscanf` returns 0 (indicating the end of the file) when you call it with a file name, Maple automatically closes the file.

The *format* specifies how Maple is to parse the input. The *format* is a Maple string made up of a sequence of conversion specifications, that may be separated by other characters. Each conversion specification has the following format, where the brackets indicate optional components.

> `% [*] [width] code`

The "`%`" symbol begins the conversion specification. The optional "`*`" indicates that Maple is to scan the object, but not return it as part of the result. It is discarded.

The optional *width* indicates the maximum number of characters to scan for this object. You can use this to scan one larger object as two smaller objects.

The *code* indicates the type of object you wish to scan. It determines the type of object that Maple returns in the resulting list. The *code* can be one of the following:

d or D   The next non-blank characters in the input must make up a signed or unsigned decimal integer. A Maple integer is returned.

o or O   The next non-blank characters in the input must make up an unsigned octal (base 8) integer. The integer is converted to decimal and returned as a Maple integer.

x or X   The next non-blank characters in the input must make up an unsigned hexadecimal (base 16) integer. The letters A through F (either capital or lower case) represent the digits corresponding to the decimal numbers 10 through 15. The integer is converted to decimal and returned as a Maple integer.

e, f, or g   The next non-blank characters in the input must make up a signed or unsigned decimal number, possibly including a decimal point, and possibly followed by E or e, an optional sign, and a decimal integer indicating a power of ten. The number is returned as a Maple floating-point value.

he, hf, or hg The following input must make up a one or two-dimensional array of floating-point (or integer) values. Characters encountered during scanning are categorized into three classes: numeric, separator, and terminator. All the characters that can appear within a number (the digits, decimal point, signs, E, e, D, and d) are numeric. Any white space, commas, or square brackets are separators. A square bracket not immediately followed by a comma, and any other character, are terminators. If a backslash is encountered, it and the following character are ignored completely.

hx The following input must make up a one or two dimensional array of floating point numbers in IEEE hex-dump format (16 characters per number). The dimensions of the array are determined as described for the "%he", "%hf", and "%hg" formats above.

s The next non-blank characters, up to but not including the following blank characters (or the end of the string), are returned as a Maple string.

a Maple collects and parses the next non-blank characters, up to but not including the following blank characters (or the end of the string). An unevaluated Maple expression is returned.

m The next characters must be a Maple expression encoded in Maple's .m file format. Maple reads enough characters to parse a single complete expression; it ignores the *width* specification. The Maple expression is returned.

c This code returns the next character (blank or otherwise) as a Maple string. If a width is specified, that many characters (blank or otherwise) are returned as a single Maple string.

[...] The characters between " [" and "] " become a list of characters that are acceptable as a character string. Maple scans characters from the input until it encounters one that is *not* in the list. The scanned characters are then returned as a Maple string.
If the list begins with a " ^ " character, the list represents all those characters *not* in the list.
If a "] " is to appear in the list, it must immediately follow the opening " [" or the " ^ " if one exists.
You can use a " - " in the list to represent a range of characters. For example, "A-Z" represents any capital letter. If a " - " is to appear as a character instead of representing a range, it must appear either at the beginning or the end of the list.

n The total number of characters scanned up to the "%n" is returned as
a Maple integer.

Maple skips non-blank characters in the *format* but not within a conversion specification (where they must match the corresponding characters
in the input). It ignores white space in the *format*, except that a space immediately preceding a "%c" specification causes the "%c" specification to
skip any blanks in the input.

If it does not successfully scan any objects, Maple returns an empty
list.

The fscanf and scanf commands use the underlying implementation
that the hardware vendor provides for the "%o" and "%x" formats. As a
result, input of octal and hexadecimal integers is subject to the restrictions
of the machine architecture.

The following example defines a Maple procedure that reads a file
containing a table of numbers, of which each row can be of a different
width. The first number in each row is an integer specifying how many
real numbers follow it in that row, and commas separate all the numbers
in each row.

```
> ReadRows := proc( fileName::string )
>    local A, count, row, num;
>    A := [];
>    do
>        # Determine how many numbers are in this row.
>        count := fscanf(fileName,"%d");
>        if count = 0 then break fi;
>        if count = [] then
>           ERROR("integer expected in file")
>        fi;
>        count := count[1];
>
>        # Read the numbers in the row.
>        row := [];
>        while count > 0 do
>           num := fscanf(fileName,",%e");
>           if num = 0 then
>              ERROR("unexpected end of file")
>           fi;
>           if num = [] then
>              ERROR("number expected in file")
>           fi;
>           row := [op(row),num[1]];
>           count := count - 1
>        od;
>
```

```
>           # Append the row to the accumulated result.
>           A := [op(A),row]
>     od;
>     A
> end:
```

Reading Maple Statements

The readstat command reads a single Maple statement from the terminal input stream. Maple parses and evaluates the statement, and returns the result. Call the readstat command as follows.

> readstat(*prompt, ditto3, ditto2, ditto1*)

The *prompt* argument specifies the prompt that readstat is to use. If you omit the *prompt* argument, Maple uses a blank prompt. You can either supply or omit all of the three arguments *ditto3, ditto2,* and *ditto1*. If you supply them, they specify the values which Maple uses for %%%, %%, and % in the statement that readstat reads. Specify each of these arguments as a Maple list containing the actual value for substitution. This allows for values that are expression sequences. For example, if % is to have the value 2*n+3 and %% is to have the value a,b, then use [2*n+3] for *ditto1* and [a,b] for *ditto2*.

The response to readstat must be a single Maple expression. The expression may span more than one input line, but readstat does not permit multiple expressions on one line. If the input contains a syntax error, readstat returns an error (which you can trap with traperror) describing the nature of the error, and its position in the input.

The following example shows a trivial use of readstat within a procedure.

```
> InteractiveDiff := proc( )
>     local a, b;
>     a := readstat("Please enter an expression: ");
>     b := readstat("Differentiate with respect to: ");
>     printf("The derivative of %a with respect to %a is %a\n",
>             a,b,diff(a,b))
> end:
```

Reading Tabular Data

The readdata command reads TEXT files containing tables of data. For simple tables, you will find this more convenient than writing your own procedure using a loop and the fscanf command.

Use the following syntax to call the readdata command.

$$\boxed{\texttt{readdata(}\ \textit{fileIdentifier},\ \textit{dataType},\ \textit{numColumns}\ \texttt{)}}$$

The *fileIdentifier* is the name or descriptor of the file from which `readdata` reads the data. The *dataType* must be one of `integer` or `float`, or you can omit it, in which case `readdata` assumes `float`. If `readdata` needs to read more than one column, you can specify the type of each column by using a list of data types.

The *numColumns* argument indicates how many columns of data are to be read from the file. If you omit *numColumns*, `readdata` reads the number of columns specified by the number of data types that you specified (one column if you did not specify any *dataType*).

If Maple reads only one column, `readdata` returns a list of the values read. If Maple reads more than one column, `readdata` returns a list of lists, each sublist of which contains the data read from one line of the file.

If you call `readdata` with a file name, and that file is not yet open, Maple opens it in READ mode as a TEXT file. Furthermore, if you call `readdata` with a file name, it automatically closes the file when `readdata` returns.

The following two examples are equivalent uses of `readdata` to read a table of (x, y, z)-triples of real numbers from a file.

```
> A1 := readdata("my_xyz_file.text",3);
```

$$A1 := [[1.5,\ 2.2,\ 3.4],\ [2.7,\ 3.4,\ 5.6],\ [1.8,\ 3.1,\ 6.7]]$$

```
> A2 := readdata("my_xyz_file.text",[float,float,float]);
```

$$A2 := [[1.5,\ 2.2,\ 3.4],\ [2.7,\ 3.4,\ 5.6],\ [1.8,\ 3.1,\ 6.7]]$$

9.6 Output Commands

Configuring Output Parameters using the `interface` Command

The `interface` command is not an output command, but you can use it to configure several parameters affecting the output produced by various commands within Maple.

To set a parameter, call the `interface` command as follows.

$$\boxed{\texttt{interface(}\ \textit{variable}\ =\ \textit{expression}\ \texttt{)}}$$

The *variable* argument specifies which parameter you wish to change, and the *expression* argument specifies the value that the parameter is to have. See the following sections or `?interface` for which parameters you can set. You may set multiple parameters by giving several arguments of the form *variable* = *expression*, with commas separating them.

To query the setting of a parameter, use the following syntax.

```
interface( variable )
```

The *variable* argument specifies the parameter to query. The `interface` command returns the current setting of the parameter. You can query only one parameter at a time.

One-Dimensional Expression Output

The `lprint` command prints Maple expressions in a one-dimensional notation very similar to the format Maple uses for input. In most cases, you could return this output to Maple as input, and the same expression would result. The single exception is if the expression contains Maple names containing non-alphanumeric characters. In that case, these names require backquotes when you supply them as input; for historical reasons, `lprint` omits the backquotes when printing such names that occur at the top level of an expression. This allows you to use `lprint` to print messages without backquotes appearing in the output; although, such usage is discouraged (you should use `printf` instead).

The `lprint` command is called as follows.

```
lprint( expressionSequence )
```

The *expressionSequence* consists of one or more Maple expressions. Each of the expressions is printed in turn, with three spaces separating each of them. Maple prints a new line character after the last expression.

Maple always sends the output that `lprint` produces to the `default` output stream. You can use the `writeto` and `appendto` commands, described later, to temporarily redirect the `default` output stream to a file.

The `interface` parameter `screenwidth` affects the output of `lprint`. If possible, Maple wraps the output between tokens. If a single token is too long to display (for example, a very long name or number), Maple breaks it across lines, and prints a backslash, "\", before each such break.

The following example illustrates `lprint` output, and how `screenwidth` affects it.

```
> lprint(expand((x+y)^5));

x^5+5*x^4*y+10*x^3*y^2+10*x^2*y^3+5*x*y^4+y^5

> interface(screenwidth=30);
> lprint(expand((x+y)^5));

x^5+5*x^4*y+10*x^3*y^2+10*x^2
```

```
*y^3+5*x*y^4+y^5
```

Two-Dimensional Expression Output

The `print` command prints Maple expressions in a two-dimensional notation. Depending on the version of Maple you are running, and the user interface you are using, this notation is either the standard math notation that appears in text books and other typeset mathematical documents, or an approximation of standard math notation using only text characters.

Use the following method to call the `print` command.

```
print( expressionSequence )
```

The *expressionSequence* consists of one or more Maple expressions. Maple prints each expression, in turn, with commas separating them.

The output produced by `print` is always sent to the `default` output stream. You can use the `writeto` and `appendto` commands, described later, to temporarily redirect the `default` output stream to a file.

Several `interface` parameters affect the output of `print`. They are set using the syntax

```
interface( parameter=value )
```

They include:

prettyprint   This selects the type of output that `print` is to produce. If you set `prettyprint` to 0, `print` produces the same output as `lprint`. If you set `prettyprint` to 1, `print` produces a simulated math notation using only text characters. If you set `prettyprint` to 2, and the version of Maple you are running is capable of it, `print` produces output using standard math notation. The default setting of `prettyprint` is 2.

indentamount   This specifies the number of spaces that Maple uses to indent the continuation of expressions that are too large to fit on a single line. This parameter takes effect only when you set `prettyprint` (see above) to 1, and/or when Maple is printing procedures. The default setting of `indentamount` is 4.

labelling or labeling   You can set this to `true` or `false`, indicating whether or not Maple should use labels to represent common subexpressions in large expressions. The use of labels can make large expressions easier to read and comprehend. The default setting of `labelling` is `true`.

labelwidth   This indicates the size that a subexpression must have in order for Maple to consider it for labeling (if `labelling` is `true`). The size is the approximate width, in characters, of the expression when printed with print and prettyprint = 1.

screenwidth This indicates the width of the screen in characters. When prettyprint is 0 or 1, Maple uses this width to decide when to wrap long expressions. When prettyprint is 2, the user interface must deal with pixels instead of characters, and determines the width automatically.

verboseproc Use this parameter when printing Maple procedures. If you set verboseproc to 1, Maple only prints user defined procedures; Maple shows system defined procedures in a simplified form giving only the arguments, and possibly a brief description of the procedure. If you set verboseproc to 2, Maple prints all procedures in full. Setting verboseproc to 3 prints all procedures in full, and prints the contents of a procedure's remember table in the form of Maple comments after the procedure.

When you use Maple interactively, it automatically displays each computed result. The format of this display is the same as if you used the print command. Therefore, all the interface parameters that affect the print command also affect the display of results.

The following example illustrates print output, and how prettyprint, indentamount, and screenwidth affect it.

```
> print(expand((x+y)^6));
```

$$x^6 + 6\,x^5\,y + 15\,x^4\,y^2 + 20\,x^3\,y^3 + 15\,x^2\,y^4 + 6\,x\,y^5 + y^6$$

```
> interface(prettyprint=1);
> print(expand((x+y)^6));

 6       5          4  2        3  3        2  4          5
x  + 6 x  y + 15 x   y  + 20 x   y  + 15 x   y  + 6 x y
          6
      + y

> interface(screenwidth=35);
> print(expand((x+y)^6));

 6       5          4  2        3  3
x  + 6 x  y + 15 x   y  + 20 x   y
            2  4         5      6
      + 15 x   y  + 6 x y  + y

> interface(indentamount=1);
> print(expand((x+y)^6));
```

```
    6     5            4 2         3 3
  x   + 6 x   y + 15 x   y  + 20 x   y
            2 4            5       6
    + 15 x   y  + 6 x y   + y
```

```
> interface(prettyprint=0);
> print(expand((x+y)^6));
```

```
x^6+6*x^5*y+15*x^4*y^2+20*x^3*y^3+
15*x^2*y^4+6*x*y^5+y^6
```

Writing Maple Strings to a File

The `writeline` command writes one or more Maple strings to a file. Each string appears on a separate line. Call the `writeline` command as follows.

> writeline(*fileIdentifier*, *stringSequence*)

The *fileIdentifier* is the name or description of the file to which you want to write, and *stringSequence* is the sequence of strings that `writeline` should write. If you omit the *stringSequence*, then `writeline` writes a blank line to the file.

Writing Arbitrary Bytes to a File

The `writebytes` command writes one or more individual characters or bytes to a file. You may specify the bytes either as a string or a list of integers.

The following syntax calls the `writebytes` command.

> writebytes(*fileIdentifier*, *bytes*)

The *fileIdentifier* is the name or descriptor of the file to which `writebytes` is writing. The *bytes* argument specifies the bytes that `writebytes` writes. This can be either a string or a list of integers. If you call `writebytes` with a file name, and that file is not yet open, Maple opens it in WRITE mode. If you specify the *bytes* as a string, Maple opens the file as a TEXT file; if you specify the *bytes* as a list of integers, Maple opens the file as a BINARY file.

The following example defines a Maple procedure which reads an entire file and copies it to a new file using `writebytes`.

```
> CopyFile := proc( sourceFile::string, destFile::string )
>    writebytes(destFile, readbytes(sourceFile, infinity));
```

```
> end:
```

Formatted Output

The `fprintf` and `printf` commands write objects to a file, using a specified format.

Call the `fprintf` and `printf` commands as follows.

```
fprintf( fileIdentifier, format, expressionSequence )
printf( format, expressionSequence )
```

The *fileIdentifier* is the name or descriptor of the file to which Maple is to write. A call to `printf` is equivalent to a call to `fprintf` with `default` as the *fileIdentifier*. If you call `fprintf` with a file name, and that file is not yet open, Maples opens it in WRITE mode as a TEXT file.

The *format* specifies how Maple is to write the elements of the *expressionSequence*. This Maple string is made up of a sequence of formatting specifications, possibly separated by other characters. Each format specification has the following syntax, where the brackets indicate optional components.

```
%[flags] [width] [.precision] code
```

The "%" symbol begins the format specification. One or more of the following flags can optionally follow the "%" symbol:

+   A signed numeric value is output with a leading "+" or "-" sign, as appropriate.

-   The output is left justified instead of right justified.

blank   A signed numeric value is output with either a leading "-" or a leading blank, depending on whether the value is negative or non-negative.

0   The output is padded on the left (between the sign and the first digit) with zeroes. If you also specify a "-", the "0" is ignored.

The optional *width* indicates the minimum number of characters to output for this field. If the formatted value has fewer characters, Maple pads it with blanks on the left (or on the right, if you specify "-").

The optional *precision* specifies the number of digits that appear after the decimal point for floating-point formats, or the maximum field width for string formats.

You may specify both *width* and/or *precision* as "∗", in which case Maple takes the *width* and/or *precision* from the argument list. The *width* and/or *precision* arguments must appear, in that order, before the argument that is being output. A negative *width* argument is equivalent to the appearance of the "−" flag.

The *code* indicates the type of object that Maple is to write. The *code* can be one of the following.

d   Formats the object as a signed decimal integer.

o   Formats the object as an unsigned octal (base 8) integer.

x or X   Formats the object as an unsigned hexadecimal (base 16) integer. Maple represents the digits corresponding to the decimal numbers 10 through 15 by the letters "A" through "F" if you use "X", or "a" through "f" if you use "x".

If the object being printed is an array of hardware floats (that is, of type hfarray), all the members are printed. The formatting width and precision are ignored, and the numbers are formatted in byte-order-independent IEEE hex dump format (16 characters wide). One-dimensional arrays are printed as one long line, with the elements separated by a space. Arrays of N dimensions, where N > 1, are printed as a sequence of N-1 dimensional arrays separated by N-1 blank lines.

e or E   Formats the object as a floating-point number in scientific notation. One digit will appear before the decimal point, and *precision* digits will appear after the decimal point (six digits if you do not specify a *precision*). This is followed by the letter "e" or "E", and a signed integer specifying a power of 10. The power of 10 will have a sign and at least three digits, with leading zeroes added if necessary.

If the object being printed is an array of hardware floats, all the members are printed. Formatting of the individual members is specified by the width and precision. Formatting of the array as a whole is as described for the "x" format above.

f   Formats the object as a fixed-point number. The number of digits specified by the *precision* will appear after the decimal point.

If the object being printed is an array of hardware floats, all the members are printed. Formatting of the individual members is specified by the width and precision. Formatting of the array as a whole is as described for the "x" format above.

g or G   Formats the object using "d", "e" (or "E" if you specified "G"), or "f" format, depending on its value. If the formatted value does not contain a decimal point, Maple uses "d" format. If the value is less than 10^{-4} or greater than $10^{precision}$, Maple uses "e" (or "E") format. Otherwise, Maple uses "f" format.

If the object being printed is an array of hardware floats, all the members are printed. Formatting of the individual members is specified by the width and precision. Formatting of the array as a whole is as described for the "x" format above.

c   Outputs the object, which must be a Maple string containing exactly one character, as a single character.

s   Outputs the object, which must be a Maple string of at least *width* characters (if specified) and at most *precision* characters (if specified).

a   Outputs the object, which can be any Maple object, in correct Maple syntax. Maple outputs at least *width* characters (if specified) and at most *precision* characters (if specified). *Note:* truncating a Maple expression by specifying a *precision* can result in an incomplete or syntactically incorrect Maple expression in the output.

m   The object, which can be any Maple object, is output in Maple's ".m" file format. Maple outputs at least *width* characters (if specified), and at most *precision* characters (if specified). *Note:* truncating a Maple ".m" format expression by specifying a *precision* can result in an incomplete or incorrect Maple expression in the output.

%   A percent symbol is output verbatim.

Maple outputs characters that are in *format* but not within a format specification verbatim.

Any of the floating-point formats can accept integer, rational, or floating-point objects; Maple converts the objects to floating-point values and outputs them appropriately.

The `fprintf` and `printf` commands do *not* automatically start a new line at the end of the output. If you require a new line, the *format* string must contain a new line character, "\n". Output from `fprintf` and `printf` is *not* subject to line wrapping at `interface(screenwidth)` characters.

The "%o", "%x", and "%X" formats use the underlying implementation that the hardware vendor provides. As a result, output of octal and hexadecimal values is subject to the restrictions of the machine architecture.

Writing Tabular Data

The `writedata` command writes tabular data to TEXT files. In many cases, this is more convenient than writing your own output procedure using a loop and the `fprintf` command.

Call the `writedata` command in the following manner.

```
writedata( fileIdentifier, data, dataType, defaultProc )
```

The *fileIdentifier* is the name or descriptor of the file to which `writedata` writes the data.

If you call `writedata` with a filename, and that file is not yet open, Maple opens it in WRITE mode as a TEXT file. Furthermore, if you call `writedata` with a file name, the file automatically closes when `writedata` returns.

The *data* must be a vector, matrix, list, or list of lists. If the *data* is a vector or list of values, `writedata` writes each value to the file on a separate line. If the *data* is a matrix or a list of lists of values, `writedata` writes each row or sublist to the file on a separate line, with tab characters separating the individual values.

The *dataType* is optional, and specifies whether `writedata` is to write the values as integers, floating-point values (the default), or strings. If you specify `integer`, the values must be numeric, and `writedata` writes them as integers (Maple truncates rational and floating-point values to integers). If you specify `float`, the values must be numeric, and `writedata` writes them as floating-point values (Maple converts integer and rational values to floating-point). If you specify `string`, the values must be strings. When writing matrices or lists of lists, you can specify the *dataType* as a list of data types, one corresponding to each column in the output.

The optional *defaultProc* argument specifies a procedure that `writedata` calls if a data value does not conform to the *dataType* you specified (for example, if `writedata` encounters a non-numeric value when the *dataType* is `float`). Maple passes the file descriptor corresponding to the *fileIdentifier*, along with the non-conforming value, as an argument to the *defaultProc*. The default *defaultProc* simply generates the error, Bad data found. A more useful *defaultProc* might be the following.

```
> UsefulDefaultProc := proc(f,x) fprintf(f,"%a",x) end:
```

This procedure is a sort of "catch-all"; it is capable of writing any kind of value to the file.

The following example computes a 5 by 5 Hilbert matrix, and writes its floating-point representation to a file.

```
> writedata("hilbertFile.txt",linalg[hilbert](5));
```

Examining the file shows:

```
1        .5       .333333 .25      .2
.5       .333333 .25      .2       .166667
.333333 .25      .2       .166667 .142857
.25      .2       .166667 .142857 .125
.2       .166667 .142857 .125     .111111
```

Flushing a Buffered File

I/O buffering may result in a delay between when you request a write operation and when Maple physically writes the data to the file. This is to capitalize on the greater efficiency of performing one large write instead of several smaller ones.

Normally, the I/O library chooses when to write to a file automatically. In some situations, however, you may desire to ensure that the data you write has actually made it into the file. For example, under UNIX, a common procedure is to run a command, such as "`tail -f` *fileName*", in another window in order to monitor the information as Maple is writing it. For cases such as these, the Maple I/O library provides the `fflush` command.

Call the `fflush` command using the following syntax.

```
fflush( fileIdentifier )
```

The *fileIdentifier* is the name or descriptor of the file whose buffer Maple is to flush. When you call `fflush`, Maple writes all information that is in the buffer but not yet in the physical file to the file. Typically, a program would call `fflush` whenever something significant is written (for example, a complete intermediate result or a few lines of output).

Note that you do not need to use `fflush`; anything you write to a file will physically be written no later than when you close the file. The `fflush` command simply forces Maple to write data on demand, so that you can monitor the progress of a file.

Redirecting the `default` Output Stream

The `writeto` and `appendto` commands redirect the `default` output stream to a file. This means that any operations that write to the `default` stream write to the file you specify instead.

You can call the `writeto` and `appendto` commands as follows.

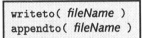

```
writeto( fileName )
appendto( fileName )
```

The *fileName* argument specifies the name of the file to which Maple is to redirect the output. If you call `writeto`, Maple truncates the file if it already exists, and writes subsequent output to the file. The `appendto` command appends to the end of the file if the file already exists. If the file you specify is already open (for example, it is in use by other file I/O operations), Maple generates an error.

The special *fileName* terminal (specified as a name, not a string) causes Maple to send subsequent default output to the original default output stream (the one that was in effect when you started Maple). The calls writeto(terminal) and appendto(terminal) are equivalent.

Issuing a writeto or appendto call directly from the Maple prompt is not the best choice of action. When writeto or appendto are in effect, Maple also writes any error messages that may result from subsequent operations to the file. Therefore, you cannot see what is happening. You should generally use the writeto and appendto commands within procedures or files of Maple commands that the read command is reading.

9.7 Conversion Commands

C or FORTRAN Generation

Maple provides commands to translate Maple expressions into two other programming languages. The languages currently supported are C and FORTRAN. Conversion to other programming languages is useful if you have used Maple's symbolic techniques to develop a numeric algorithm, which then may run faster as a C or FORTRAN program than as a Maple procedure.

Perform a conversion to FORTRAN or C using the fortran or C commands, respectively. There are also several support commands for code generation, which can be found in the codegen package. Because C is a simple name which you may likely use as a variable, Maple does not preload the C command, and so you must load it using readlib first.

Call the fortran and C commands using the following syntax.

```
fortran( expression, options )
C( expression, options )
```

The *expression* can take one of the following forms:

1. A single algebraic expression: Maple generates a sequence of C or FORTRAN statements to compute the value of this expression.

2. A list of expressions of the form *name= expression*: Maple generates a sequence of statements to compute each *expression* and assigns it to the corresponding *name*.

3. A named array of expressions: Maple generates a sequence of C or FORTRAN statements to compute each expression and assign it to the corresponding element of the array.

4. A Maple procedure: Maple generates a C function or FORTRAN subroutine.

The `fortran` command uses the `'fortran/function_name'` command when translating function names to their FORTRAN equivalents. This command takes three arguments: the Maple function name, the number of arguments, and the precision, and returns a single FORTRAN function name. You can override the default translations by assigning values to the remember table of `'fortran/function_name'`.

> `'fortran/function_name'(arctan,1,double) := datan;`

$$\text{‘fortran/function\_name‘(arctan, 1, }double) := datan$$

> `'fortran/function_name'(arctan,2,single) := atan2;`

$$\text{‘fortran/function\_name‘(arctan, 2, }single) := atan2$$

When translating arrays, the C command re-indexes all array indices to begin with 0, since the base of C arrays is 0. The `fortran` command re-indexes arrays to begin with 1, but only when Maple is translating a procedure.

Here Maple symbolically calculates the anti-derivative.

> `f := unapply( int( 1/(1+x^4), x), x );`

$$f := x \rightarrow \frac{1}{8}\sqrt{2}\ln\left(\frac{x^2 + x\sqrt{2} + 1}{x^2 - x\sqrt{2} + 1}\right) + \frac{1}{4}\sqrt{2}\arctan(x\sqrt{2} + 1)$$

$$+ \frac{1}{4}\sqrt{2}\arctan(x\sqrt{2} - 1)$$

The `fortran` command generates a FORTRAN routine.

> `fortran(f, optimized);`

```
c The options were    : operatorarrow
      real function f(x)
      real x
      real t1
      real t12
      real t16
      real t2
      real t3
      real t8
        t1 = sqrt(2.E0)
        t2 = x**2
        t3 = x*t1
        t8 = alog((t2+t3+1)/(t2-t3+1))
        t12 = atan(t3+1)
        t16 = atan(t3-1)
```

```
        f = t1*t8/8+t1*t12/4+t1*t16/4
      return
      end
```

Now translate the same expression to C.

```
> readlib(C):
> C(f, optimized);


/* The options were      : operatorarrow */
double f(x)
double x;
{
  double t2;
  double t3;
  double t8;
  double t16;
  double t1;
  double t12;
  {
    t1 = sqrt(2.0);
    t2 = x*x;
    t3 = x*t1;
    t8 = log((t2+t3+1.0)/(t2-t3+1.0));
    t12 = atan(t3+1.0);
    t16 = atan(t3-1.0);
    return(t1*t8/8.0+t1*t12/4.0+t1*t16/4.0);
  }
}
```

LaTeX or *eqn* Generation

Maple supports conversion of Maple expressions to two typesetting languages: LaTeX and *eqn*. Conversion to typesetting languages is useful when you need to insert a result in a scientific paper.

You can perform conversion to LaTeX and *eqn* using the latex and eqn commands. Because eqn is a simple name which you may likely use as a variable, Maple does not preload the eqn command, and so you must load it using readlib first.

Call the latex and eqn commands as follows.

```
latex( expression, fileName )
eqn( expression, fileName )
```

The *expression* can be any mathematical expression. Maple-specific expressions, such as procedures, are not translatable. The *fileName* is optional, and specifies that Maple writes the translated output to the file you specified. If you do not specify a *fileName*, Maple writes the output to the default output stream (your session).

The `latex` and `eqn` commands know how to translate most types of mathematical expressions, including integrals, limits, sums, products, and matrices. You can expand the capabilities of `latex` and `eqn` by defining procedures with names of the form ‘latex/*functionName*‘ or ‘eqn/*functionName*‘. Such a procedure is responsible for formatting calls to the function called *functionName*. You should produce the output of such formatting functions with `printf`; `latex` and `eqn` use `writeto` to redirect the output when you specify a *fileName*.

Neither command generates the commands that LATEX or *eqn* require to put the typesetting system into mathematics mode ($...$ for LATEX, and .EQ....EN for *eqn*).

The following example shows the generation of LATEX and *eqn* for an integral and its value. Notice the use of Int, the inert form of int, to prevent evaluation of the left hand side of the equation that Maple is formatting.

```
> Int(1/(x^4+1),x) = int(1/(x^4+1),x);
```

$$\int \frac{1}{x^4+1}\, dx = \frac{1}{8}\,\sqrt{2}\ln\left(\frac{x^2+x\sqrt{2}+1}{x^2-x\sqrt{2}+1}\right)$$

$$+\frac{1}{4}\,\sqrt{2}\,\arctan(x\sqrt{2}+1)+\frac{1}{4}\,\sqrt{2}\,\arctan(x\sqrt{2}-1)$$

```
> latex(%);
\int \!\left ({x}^{4}+1\right )^{-1}{dx}=1/8\,
\sqrt {2}\ln ({\frac {{x}^{2}+x\sqrt {2}+1}{{x
}^{2}-x\sqrt {2}+1}})+1/4\,\sqrt {2}\arctan(x
\sqrt {2}+1)+1/4\,\sqrt {2}\arctan(x\sqrt {2}-
1)
```

```
> readlib(eqn):
> eqn(%%);
{{int  { {( {{  "x"   sup 4 }^+^1 } )} sup -1 }~d   "x"
}~~=~~{{ {{ sqrt 2 }^{ln ( { {{  "x"   sup 2 }^+^{ "x"
^{ sqrt 2 }}^+^1 } over {{ "x"   sup 2 }^-^{ "x" ^{ sqrt
2 }}^+^1 }})}}} over 8 }^+^{ {{ sqrt 2 }^ {arctan ( {{
"x" ^{ sqrt 2 }}^+^1 })}} over 4 }^+^{ {{ sqrt 2 }^{
```

```
arctan ( {{ "x" ^{ sqrt 2 }}^-^1 })}} over 4 }}}
```

Conversion between Strings and Lists of Integers

The readbytes and writebytes commands described in *Reading Arbitrary Bytes from a File* on page 341 and *Writing Arbitrary Bytes to a File* on page 350 can work with either Maple strings or lists of integers. You can use the convert command to convert between these two formats as follows.

```
convert( string, bytes )
convert( integerList, bytes )
```

If you pass convert(...,bytes) a string, it returns a list of integers; if you pass it a list of integers, it returns a string.

Due to the way strings are implemented in Maple, the character corresponding to the byte-value 0 cannot appear in a string. Therefore, if *integerList* contains a zero, convert returns a string of only those characters corresponding to the integers preceding the occurrence of 0 in the list.

Conversion between strings and lists of integers is useful when Maple must interpret parts of a stream of bytes as a character string, while it must interpret other parts as individual bytes.

In the following example, Maple converts a string to a list of integers. Then, it converts the same list, but with one entry changed to 0, back to a string. Notice that the string is truncated at the location of the 0.

```
> convert("Test String",bytes);
```

$$[84, 101, 115, 116, 32, 83, 116, 114, 105, 110, 103]$$

```
> convert([84,101,115,116,0,83,116,114,105,110,103],bytes);
```

"Test"

Parsing Maple Expressions and Statements

The parse command converts a string of valid Maple input into the corresponding Maple expression. The expression is simplified, but not evaluated.

Use the parse command as follows.

```
parse( string, options )
```

The *string* argument is the string that needs parsing. It must describe a Maple expression (or statement, see below) using the Maple language syntax.

You may supply one or more *options* to the parse command:

statement   This indicates that parse is to accept statements in addition to expressions. However, since Maple does not allow the existence of unevaluated statements, parse does evaluate the *string* if you specify statement.

nosemicolon   Normally, parse supplies a terminating semicolon, ";" if the string does not end in a semicolon or a colon, ":". If you specify nosemicolon, this does not happen, and Maple generates an unexpected end of input error if the string is incomplete. The readstat command, which uses readline and parse, makes use of this facility to allow multi-line inputs.

If the *string* passed to parse contains a syntax error, parse generates an error (which you can trap with traperror) of the following form.

```
incorrect syntax in parse:
errorDescription (errorLocation)
```

The *errorDescription* describes the nature of the error (for example, '+' unexpected, or unexpected end of input). The *errorLocation* gives the approximate character position within the string at which Maple detected the error.

When you call parse from the Maple prompt, Maple displays the parsed result depending on whether the call to parse ends in a semicolon or a colon. Whether the *string* passed to parse ends in a semicolon or a colon does not matter.

```
> parse("a+2+b+3");
```

$$a + 5 + b$$

```
> parse("sin(3.0)"):
> %;
```

$$.1411200081$$

Formatted Conversion to and from Strings

The sprintf and sscanf commands are similar to fprintf/printf and fscanf/scanf, except that they read from or write to Maple strings instead of files.

Call the sprintf command using the following syntax.

> ```
> sprintf( format, expressionSequence )
> ```

The *format* specifies how Maple is to format the elements of the *expressionSequence*. This Maple string is made up of a sequence of formatting specifications, possibly separated by other characters. See *Formatted Output* on page 351.

The sprintf command returns a string containing the formatted result. Call the sscanf command as follows.

> ```
> sscanf( sourceString, format )
> ```

The *sourceString* provides the input for scanning. The *format* specifies how Maple is to parse the input. A sequence of conversion specifications (and possibly other anticipated characters) make up this Maple string. See *Formatted Input* on page 341. The sscanf command returns a list of the scanned objects, just as fscanf and scanf do.

The following example illustrates sprintf and sscanf by converting a floating-point number and two algebraic expressions into a floating-point format, Maple syntax, and Maple .m format, respectively. This string is then parsed back into the corresponding objects using sscanf.

```
> s := sprintf("%4.2f %a %m",evalf(Pi),sin(3),cos(3));
```
$$s := \text{``3.14 sin(3) -\%\$cosG6\#\textbackslash\text{``}\textbackslash \text{``\$''}}$$

```
> sscanf(s,"%f %a %m");
```
$$[3.14, \sin(3), \cos(3)]$$

9.8 A Detailed Example

This section provides an example that uses several of the I/O facilities described in this chapter to generate a FORTRAN subroutine in a text file. In this example, you can find all of the required Maple commands typed on the command line. In general, for such a task you would write a procedure or, at the very least, a file of Maple commands.

Suppose you wish to compute values of the function $1 - \text{erf}(x) + \exp(-x)$ for many points on the interval [0,2], accurate to five decimal places. Using the numapprox package from the Maple library, you can obtain a rational approximation for this function as follows.

```
> f := 1 - erf(x) + exp(-x):
> approx := numapprox[minimax](f, x=0..2, [5,5]);
```

$$approx := (1.872591139 + (-2.480795187+$$

$$(1.455363450 + (-.4104063767 + .04512839694\,x)\,x)\,x$$

$$)x)/(.9362966183 + (-.2440899305+$$

$$(.2351120169 + (.00114989244 - .01091415865\,x)\,x)\,x$$

$$)x)$$

You can now create the file and write the subroutine header to the file.

```
> file := "approx.f77":
> fprintf(file, "real function f(x)\nreal x\n"):
```

Before you can write the actual FORTRAN output to the file, you must close the file. Otherwise, the fortran command attempts to open the file in APPEND mode, which results in an error if the file is already open.

```
> fclose(file):
```

Now you can write the actual FORTRAN statements to the file.

```
> fortran(['f'=approx], filename=file):
```

Finally, you add the remainder of the FORTRAN subroutine syntax.

```
> fopen(file, APPEND):
> fprintf(file, "return\nend\n"):
> fclose(file):
```

If you now examine the file, it looks like this:

```
real function f(x)
real x
     f = (0.187258E1+(-0.2480777E1+(0.1455351E1+
#(-0.4104024E0+0.4512788E-1*x)*x)*x)*x)/(0.9
#362913E0+(-0.2440864E0+(0.235111E0+(0.11504
#53E-2-0.1091373E-1*x)*x)*x)*x)
return
end
```

This subroutine is now ready for you to compile and link into a FORTRAN program.

9.9 Notes to C Programmers

If you have experience programming in the C or C++ programming languages, many of the I/O commands described in this chapter will seem familiar. This is not coincidental, as the Maple I/O library design purposely emulates the C standard I/O library.

In general, the Maple I/O commands work in a similar manner to their C counterparts. The differences that arise are the result of differences between the Maple and C languages, and how you use them. For example, in the C library, you must pass the `sprintf` function a buffer into which it writes the result. In Maple, strings are objects that you can pass around as easily as numbers, so the `sprintf` command simply returns a string that is sufficiently long to hold the result. This method is both easier to work with and less error prone, as it removes the danger of writing past the end of a fixed length buffer.

Similarly, the `fscanf`, `scanf`, and `sscanf` commands return a list of the parsed results instead of requiring you to pass references to variables. This method is also less error prone, as it removes any danger of passing the wrong type of variable or one of insufficient size.

Other differences include the use of a single command, `filepos`, to perform the work of two C functions, `ftell` and `fseek`. You can do this in Maple, since functions can take a variable number of arguments.

In general, if you have C or C++ programming experience, you should have very little trouble using the Maple I/O library.

9.10 Conclusion

This chapter has revealed the details of importing and exporting data and code into and out of Maple. Most commands discussed in this chapter are more primitive than those commands which you are likely to use, such as `save` and `writeto`. The aforementioned Maple commands ensure that you are properly equipped to write specialized exporting and importing procedures. Their basis is similar to the commands found in the popular C programming language, although they have been extended to allow easy printing of algebraic expressions.

Overall, this book provides an essential framework for understanding Maple's programming language. Each chapter is designed to teach you to use a particular area of Maple effectively. However, a complete discussion of Maple can not fit into a single book. The Maple help system is an excellent resource and complements this volume. While this book teaches fundamental concepts and provides a pedagogical introduction to topics, the help system provides the details on each command and feature.

Also, numerous authors have published many books about Maple. These include not only books, such as this one, on the general use of Maple, but also books directed toward the use of Maple in a particular field or application. Should you wish to consult books that parallel your own area of interest, this book will still serve as a handy reference and guide to Maple programming.

Index

of ranges, 166
of relations, 158
of sequences, 59, 142
of series, 164
of sets, 145
of strings, 135
of tables, 163
of unevaluates expressions, 167
set of, 60
structured, 60, 90, 169
with special evaluation, 58
typesetting, 358

unapply, 76, 265
unassigning, 121
unbuffered files, 334
undefined, 299, 304, 309, 310
underscores, 111
uneval, 58, 59, 168, 242
unevaluated calls, 105
unevaluated expressions, 167, 242
unevaluated returns, 14
uniform, 54–56
uniform distribution, 54
uniform random number generator, 54
union, 91, 110
unsigned floats, 137
unstopat, 215, 220, 221
unstoperror, 224
unstopwhen, 223
user input, 86, 88
userinfo, 97

v, 50
var, 247, 312
variables
environment, 48

global, 6, 80, 81, 187
identifying, 79
local, 4, 6, 79, 85, 187
scope of, 6, 50
unassigning, 81, 121
undeclared, 51
varyAspect, 315
vector fields, 297
vectorfieldplot, 301–307
vectors
merging, 175
verboseproc, 6, 193, 349
vertices, 64
VIEW, 270

watchpoints, 215, 222
and error messages,
*error watchpoints*223
removing, 216, 223
setting, 222
showing, 231
whattype, 59
where, 228–230
while, 10, 18–20, 29, 110, 126–128,
159, 161, 162, 172, 236
default value, 125
white space, 113
with, 98–100, 102, 104, 206
WRITE, 335–337, 339, 350, 351, 354
writebytes, 350, 360
writedata, 332, 353, 354
writeline, 350
writeto, 347, 348, 355, 356, 359, 364

yellowsides, 275

zip, 172, 175